Environmental Ethics and Sustainable Design

Through the introduction of BioDialectics, a conceptual approach intended to extend the ethical considerations of sustainable design, this book analyses how the field engages with and adapts ideas from environmental ethics.

Sustainable design is often presented as a response to environmental crises from a socio-technical perspective, while its ethical foundations remain relatively underexamined. To address this issue, the book systematically examines sustainable design ethics through two broad orientations: anthropocentric and biocentric approaches. Rather than treating these perspectives as mutually exclusive, BioDialectics moves beyond traditional biocentrism's emphasis on the intrinsic value of all living beings by incorporating relational and systemic understandings of human–nature relations. By taking into account historical, geopolitical, and structural factors that shape these relations, BioDialectics proposes an expanded conception of justice that includes both ecological and social concerns and considers the environmental burdens experienced by ecosystems and communities. Drawing on classical and contemporary philosophical work, as well as examples from sustainable design education and sustainable living initiatives, the book provides a structured analysis and conceptual vocabulary for sustainable design ethics. Each chapter identifies key challenges and possibilities, encouraging reflection on the roles and responsibilities of designers.

The book will be of interest to academics, researchers, and practitioners working at the intersection of ethics, environment, and sustainable design.

Martina Maria Keitsch is Professor in Sustainable Design at the Department of Design, Norwegian University of Science and Technology, Trondheim, Norway. She holds a Ph.D. in Philosophy with a specialization in environmental ethics. Currently she is Ph.D. Programme Leader at the Department of Design and Head of the Ph.D. Council at the Faculty of Architecture and Design. Keitsch has worked for over 25 years in the sustainable development area, mainly with eco-design, architecture, urban planning, stakeholder inclusion and ethics of sustainability.

Routledge Studies in Sustainability

Transdisciplinary Research, Sustainability, and Social Transformation
Governance and Knowledge Co-Production
Tom Dedeurwaerdere

Digital Technologies for Sustainable Futures
Promises and Pitfalls
Edited by Chiara Certomà, Fabio Iapaolo, and Federico Martellozzo

Rethinking the Concept of Waste and Mass Consumption
Preserving Resources through Reuse, Repair and Recycling
Richard Waite

Mediating Sustainability in the Consumer Society
Edited by Astrid Skjerven, Lisbeth Løvbak Berg, Liv Merete Nielsen, and Dagny Stuedahl

Smart Cities to Smart Societies
Moving Beyond Technology
Edited by Esmat Zaidan, Imad Antoine Ibrahim, and Elie Azar

Events of Art and Education in Post-climate Times
Carl Anders Säfström and Glenn Loughran

Sustainability Perception Indicators
Theory and Practice of Perception Indicators in Sustainability
Stephen Morse

Environmental Ethics and Sustainable Design
Reflections on Theory and Practice
Martina Maria Keitsch

For more information on this series, please visit: www.routledge.com/Routledge-Studies-in-Sustainability/book-series/RSSTY

Environmental Ethics and Sustainable Design

Reflections on Theory and Practice

Martina Maria Keitsch

Routledge
Taylor & Francis Group
LONDON AND NEW YORK

First published 2026
by Routledge
4 Park Square, Milton Park, Abingdon, Oxon OX14 4RN

and by Routledge
605 Third Avenue, New York, NY 10158

Routledge is an imprint of the Taylor & Francis Group, an informa business

British Library Cataloguing-in-Publication Data
A catalogue record for this book is available from the British Library

ISBN: 978-1-032-46986-7 (hbk)
ISBN: 978-1-032-46991-1 (pbk)
ISBN: 978-1-003-38409-0 (ebk)

DOI: 10.4324/9781003384090

Typeset in Times New Roman
by Taylor & Francis Books

Contents

Thank You from the Author

I express my sincere gratitude to all those who generously gave their time and shared their stories throughout the course of this project. Your patience, openness, and generosity made this work possible.

A special thanks goes to Helmut Herzberg for his unique insights and unwavering support to explore and discuss new ideas. His contributions deeply enriched this book.

Foreword

> "The result of philosophy is not a number of 'philosophical propositions', but to make propositions clear".
>
> Wittgenstein (1922, 10)

This book explores the intersections of environmental ethics and sustainable design in both theory and practice, with particular attention to their interrelations. Rather than offering prescriptive solutions, it aims to open new pathways for thinking about sustainability, ethics, and design.

It is intended for researchers and postgraduate students in architecture, design, social sciences, engineering, and sustainability studies. It may also interest educators, policymakers, and institutional planners engaged in transdisciplinary collaboration for sustainable development.

Given its interdisciplinary and reflective nature, the book invites a rhizomatic reading that moves fluidly between chapters and revisits key ideas. Its most significant contribution is the development of *BioDialectics*, a novel concept that reframes sustainable design through ecological interconnectedness, post-anthropocentric perspectives, and epistemic justice.

Ultimately, this book serves as an intellectual companion rather than a practical manual, prioritising inquiry over definitive answers. Readers are invited to engage with its themes, cultivating their own space for philosophical reflection and critical questioning.

Reference

Wittgenstein, L., (1922). *Tractatus Logico-Philosophicus.* Routledge & Kegan Paul

1 Introduction

1.1 Setting the Scene: Environmental Ethics and Sustainable Design

It is 2025, and the ecological crisis has become a global threat, evident in melting glaciers, rising sea levels, and increasingly extreme temperatures. Beyond endangering ecosystems, its effects have profound social and ethical impacts, demanding a larger debate on societal values such as respect for nature, justice, and fairness.[1] The following chapter looks at the development of environmental ethics (EE) and sustainable design (SD) and their reflections on the environmental crisis.

Environmental concerns have permeated Western culture from antiquity to modernity. For example, Plato (c. 360 BC) critiques resource exploitation in *Kritias*, highlighting deforestation and soil erosion as consequences of human intervention. *"Many great deluges have taken place during the nine thousand years, for that is the number of years which have elapsed since the time of which I am speaking; and during all this time and through so many changes, there has never been any considerable accumulation of the soil coming down from the mountains, as in other places, but the earth has fallen away all round and sunk out of sight. The consequence is, that in comparison of what then was, there are remaining only the bones of the wasted body, as they may be called, as in the case of small islands, all the richer and softer parts of the soil having fallen away, and the mere skeleton of the land being left"*. Seneca (63CE) predicts disastrous consequences of urban sprawl, and Malthus (1798) points to the chasm between excessive human reproduction and lack of resources in modern societies.

Since the late 1960s, the term has connoted environmental deterioration on a global scale. Published in 1962, Rachel Carson's *Silent Spring* was one of the first works that introduced environmental concerns to the general public. Carson illustrates how the biosphere is compromised by synthetic pesticides that are eventually poisoning humans by moving up the food chain. Being primarily a scientist, Carson's criticism of the capitalist system[2] also generated, however, awareness of interdependence of environmental and societal challenges in the form of undermining fundamental rights, deepening inequalities, and contributing to ecological marginalisation (Kousis, 2002).

DOI: 10.4324/9781003384090-1

The first publication to explicitly mention the term "ecological crisis" appeared in White (1967), followed by Ehrlich's *The Population Bomb* (1968) and Meadows et al.'s *Limits to Growth* (1972), both of which further challenged the logic of capitalist economies and their commitment to infinite production and consumption. In the final decade of the 20th century, scientists, researchers, and policymakers increasingly acknowledged key future environmental challenges: biodiversity loss, stratospheric ozone depletion, climate change, and the collapse of natural resource stocks. In its 1987 report *Our Common Future*, the World Commission on Environment and Development articulated a global response to these issues through the concept of "sustainable development", which became a central concept in the decades that followed.

Environmental ethics emerged as a distinct philosophical discipline in the early 1970s. It is regarded as a branch of bioethics, which also encompasses medical ethics, animal ethics, genetic ethics, and population ethics (Fiore, 2016). EE examines moral principles, values, duties[3] and responsibilities that humans hold towards nature and non-human beings, while two distinct forms of moral concern have been prevalent in the last decades: biocentrism (including eco-centrism and organicism), and anthropocentrism.

Biocentrism aims at protection of the environment for providing the wellbeing of living beings or nature as a whole. Anthropocentrism aims to protect the environment for providing the wellbeing of humans. The epistemic[4] background for biocentrism is the assumption of a biotic community as an interdependent, functioning system. The evolution of life in this system is represented through various forms, while the simplest forms of beings are still parts of the most complex ones. Even the most complex beings, i.e. humans, cannot exist outside of this system. In this respect, humans are part of nature; they did not create it but have evolved alongside all other natural beings.

Biocentric core *values* are intrinsic[5], grounded in the belief that all living beings possess self-value and/or self-purpose ("auto-telos"). Self-purpose, the intrinsic aim of all living beings to grow and realise their natural purpose, is a key biocentric argument for ascribing moral value to non-human life, as articulated for example by Rolston (2006):

> *"Values are intrinsic, instrumental, and systemic, and all three are interwoven. It would be foolish to value the golden eggs and disvalue the goose that lays them". It would be a mistake to value the goose only instrumentally. A goose that lays golden eggs is systemically valuable. How much more so is an ecosystem that generates myriads of species; or even, as we next see, an Earth that produces billions of species, ourselves included".*
>
> (Ibid., 8)

The central maxim of biocentrism based on which beings or systems are recognised as moral subjects is that humans have a duty to protect and preserve nature. Biocentrism encompasses deontological and teleological positions. The deontological view derives duties from the existence of living beings, ecosystems, or nature as a whole, while the teleological view derives duties from the purposiveness or inherent interests of those beings and systems (Hauskeller, 2005). Individual positions in biocentrism and anthropocentrism are discussed in detail in Chapter 2.

The epistemic background for anthropocentrism is the assumption of nature as human habitat, shaped and maintained according to humans' needs and desires. This position implies a "hetero-telos" ("other-purpose", i.e. purpose assigned by others, opposed to "self-purpose") of nature and living beings as a basis for teleological or deontological arguments.

Teleological anthropocentrism refers to human interests and ascribes value to nature or other living beings in meeting those interests. Duties towards nature exist only insofar as something has an instrumental value as a resource, or non-instrumental contemplative, theoretical, religious, socio-cultural or aesthetic value. Deontological anthropocentrism ascribes only indirect value and duties to nature. Compassionate and protective actions toward natural beings, however, can contribute to an individual's moral development for example, by advancing forms of care or love that are not motivated by personal gain. In terms of justice, the destruction of nature interferes with human rights, e.g., by depriving future generations of having access to nature and experiences. In either position, anthropocentrism subscribes to a resource conservation ethics.

Although environmental ethics is a comparatively recent development in the history of philosophy, it has triggered a foundational debate in moral theory on whether moral rights should be based exclusively on human abilities such as rationality, reasoning, and decision-making, or if nature should have rights as well and on which premises. The growing global relevance of sustainable development has also contributed to making environmental ethics more approachable and relevant to a wider range of audiences. EE concepts have for example increasingly permeated academic disciplines such as design, urban planning, and architecture, and are being integrated into both theoretical structures and practical applications.

Ideas that emphasise justice, fairness, and human well-being are foundational to many design frameworks. Within these frameworks, sustainable design reflects values and norms concerning human–nature relationships, inspiring ethical discourses on nature, just patterns of production and consumption, and professional responsibilities embedded within design practices. At the same time, design thinking frequently comprises dialectical tensions, as Swann (2002) underlines in an historical analysis of the discipline, explaining that from the early 20th century onwards, design underwent a process of increasing commercialisation.

By the 1960s and 1970s, industrial design had evolved into a prominent marketing strategy and became increasingly aligned with engineering disciplines. Epistemologically, early industrial design was situated within a positivist paradigm, privileging empirical methods and problem-solving orientation. This orientation often paralleled maxims of social systems engineering, such as the belief that an "objective" professional could inscribe economically or ecologically "desirable meaning" into artefacts through the design process. However, the notion of professional objectivity has been the subject of sustained critique across multiple disciplines, most notably by Habermas (1987) who disputes that engineers and designers are disembodied, value-neutral agents but asserts that they are socially and environmentally situated actors. When design is steered predominantly by technical and instrumental rationality, it risks circumventing participatory processes, marginalising plural epistemologies, and uncritically reproducing dominant socio-economic values within material culture.

From a societal perspective, the *colonisation of the lifeworld* emerges when the communicative domains of everyday life become increasingly permeated by technical and bureaucratic rationality:

> *"The communicative practice of everyday life is one-sidedly rationalised into a utilitarian lifestyle; this media-induced shift to purposive-rational action orientations calls forth the reaction of a hedonism freed from the pressures of rationality. The bureaucratic disempowering and desiccation of spontaneous processes of opinion- and will-formation expands the scope for engineering mass loyalty and makes it easier to uncouple political decision-making from concrete, identity-forming contexts of life".*
>
> (Habermas 1987, 325)

As industrial production progressively became the natural habitat for design from the 1960s onwards, critical voices against the dominance of commercial interests and marketing-driven design imperatives appeared, and they often intersected with growing ecological concerns. In 1964, a group of graphic designers published the "First Things First" manifesto, in which they advocated a reorientation of design practice toward socially meaningful and environmentally responsible purposes, rather than serving the interests of consumerism alone, "a reversal of priorities in favour of more useful, lasting and democratic forms of communication – a mind shift away from product marketing and towards the exploration and production of a new kind of meaning" (Garland, 2000). In 1972, Papanek, an early proponent of the moral and societal responsibility of designers, cautioned that: "As a politically dis-interested but morally motivated animal, the designer should not be a … pimp for the excesses of big business interests" (1991, 327). Around the same time, ecological theory began to influence design thinking as well. The Gaia hypothesis, formulated by Lovelock and Margulis (1974), proposed that the Earth functions as a self-regulating system maintained through the interactions of all living organisms. This vision inspired designers to reflect more deeply on the interdependence of industrial culture and natural systems, and to explore more ecologically attuned forms of design practice.

Since the 1980s, the design disciplines have undergone significant transformations, shifting for example from a primary focus on creating consumer products and services to the adoption of user-centred approaches. Contemporary design concepts increasingly emphasise emotion, experience, and meaningful human interaction (Norman, 2004), highlight the importance of experiential qualities (Hassenzahl, 2003), and promote active user involvement in the design process (Sanders, 2002). In parallel, sustainable design and planning concepts have moved beyond technical solutions, seeking to address diverse stakeholder needs and values, and integrating responsibility, social well-being, and inclusion (Manzini, 2006). In some cases, design ideas align with activist-oriented agendas that call for a more politically engaged role for designers, consumers, and stakeholders (Fuad-Luke, 2009).

A prevailing credo within sustainable design discourses is that future-oriented practices should not only address the "triple bottom line" – economic viability, environmental responsibility, and social equity – but also cultivate new modes of relating to the environment, ultimately reconciling human with nature relationships. These aspirational goals warrant a systematic ethical analysis of sustainable design concepts, a task undertaken in Chapters 2 and 3. As part of the broader inquiry into the ethical foundations of sustainable design, the following excursion examines how evolving human conceptions of nature have informed environmental philosophy across intellectual history.

1.2 Excursion: Changing Nature Perceptions

The term *nature* is derived from the Greek *physis* ("to blossom") and the Latin *nasci* ("to arise") (Du Carme et al., 2020), and it denotes that which develops independently. Since antiquity, natural phenomena (*natura naturata* [6]) are contrasted to the social world and to human-made artefacts. Humans rarely encounter "genuine" nature, as they usually inhabit culturally shaped environments. Yet, the embodied experience of one's own physicality (inner nature) and the surrounding environment (outer nature), however temporary, contingent, and fragmented in its individual manifestations, allows us to comprehend nature dynamics (*natura naturans*). It further reveals that nature precedes human existence, resists complete control, and persists beyond the existence of the human species.

Reflecting on the relationship between humans and nature belongs to a longstanding philosophical tradition. In the *Nicomachean Ethics* (c. 350 BCE), Aristotle first introduces the triad of the truth, the good, and the beautiful, which guides both natural processes and human conduct. Nature (*physis*), society (*praxis*), and technology (*poiesis*) each correspond to distinct dimensions of human faculties and modes of reasoning. *Physis*, associated with truth, aligns with the domain of *theoria* and engages the faculty of theoretical reason *epistēmē* (scientific knowledge) and *sophia* (philosophical wisdom). This mode seeks to understand the cosmos and the natural world through contemplation and inquiry. *Praxis* relates to the pursuit of the good, both at the individual and societal levels. It involves the faculty of moral intelligence, or *phronēsis*, which guides ethical action and decision-making aimed at *eudaimonia* [7] human flourishing. Finally, *poiesis* pertains to the domain of technē (art and craft), expressing the human capacity to make and produce. It is linked to the creative transformation of materials and environments, often through technology, with ends that are external to the act itself. (Keitsch, 2006).

Beyond its ethical foundations, Aristotle's concept of telos[8] represents a core principle in biocentrism, supporting the recognition of the intrinsic value and directedness of all forms of life. Purpose (telos) is represented in nature ("physis" in singularity, and "cosmos" in totality), as well as in moral action (praxis) and in artificial works (poiesis). Aristotle's concept of "entelecheia" – entelechy, intrinsic form – constitutes the most important aspect for any purposefulness, as for example described in his "Physics":

> *"Now intelligent action is for the sake of an end; therefore, the nature of things also is so. Thus, if a house, e.g., had been a thing made by nature, it would have been made in the same way as it is now by art; and if things made by nature were made also by art, they would come to be in the same way as by nature. Each step then in the series is for the sake of the next; and generally, art partly completes what nature cannot bring to a finish and partly imitates her. If, therefore, artificial products are for the sake of an end, so clearly also are natural products".*
>
> (Aristotle, n.d. Book II, Part 8)

For Aristotle, nature is superior to technology because it produces autonomously, whereas technology is always allopoietic, i.e. created by someone. Technology (*poiesis*) holds little significance for either contemplative understanding (*theoria*) or moral action (*praxis*) (Aristotle, 2009). While he agrees that technological knowledge stems from observing the phenomenal realm, Aristotle's philosophy contains no concept of "applied" science. The highest human purpose and the best human life are realised through a morally impeccable and materially frugal contemplative life (*bios theoretikos*). However, the well-being of a community depends on the willingness of some individuals to lead a second-best life devoted to politics (*bios politikos*).

Interpretations of nature change with early Christianity, becoming a harbinger of many modern ideas. Among others, the view of a poetic (productive) nature is replaced by the idea of the Christian god as Creator, who, ex nihilo, brings forth all forms and beings.[9] An anticipated imperfection and sinfulness of the material world (and the human body) render the phenomenological world deceitful and vacant. Morality consists of obeying God's laws, doing good deeds, and abstaining from fleshly desires. In the High Medieval period (500–1500 CE), urban culture in Europe and increased trade between countries and continents began to flourish. Yet it was not until the 13th century, with the renaissance of Aristotelian philosophy, alongside the reception of Augustinian ideas, that a revised view of the relationship between humans and nature emerged.

Augustine's existentialist turn[10] states that humans achieve knowledge through looking within: *"Do not go outside; return to yourself. In our interior the truth resides"* (Augustine 2005, 72). Augustine initiates a paradigm shift in Christian philosophy, stating that knowledge and truth coming from God are accessible through self-reflection. He also perceives that everything living is a product of God and that living beings partially participate in God's perfection. This transforms early Christianity's perception of nature, from a mute vale of tears to a visible pendant of God, accessible to human faith and contemplation.

In the mid-13th century, Aquinas (1920) reintroduces the idea of *lumen naturale* ("natural light", for Aquinas, "natural light of reason"), stating that knowledge of natural phenomena is generated when reason develops ideas from sensory impressions. Ontologically, Aquinas distinguishes between form and matter, and between reason and the senses. Forms are the essence of things and originally exist as ideas in the mind of God. Since God is supreme, each realm of reality is considered higher the more it consists of form or idea and lower the more it consists of matter. There is a continuous progression from plant and animal life to the rational human soul, and further into the realm of pure spirits – who also govern the stars – ultimately leading to God:

> *"the operation of grace proceeds in orderly fashion no less than the operation of nature. Now nature always makes a beginning with perfect things … Therefore, it seems that the operation of grace also began with perfect things, so that those who were the first to deliver the faith, knew it most perfectly".*
>
> (Aquinas (1920, Second part of the second part. Question 1, Article 7, objection 3).

In 1543, Nicolaus Copernicus published *On the Revolutions of the Celestial Spheres*, asserting that the Earth and other planets revolve around the Sun. This heliocentric model marked the beginning of a profound transformation in scientific thought. In 1609, Johannes Kepler's *A New Astronomy* further advanced this transformation by demonstrating that planetary orbits are elliptical rather than circular. These insights profoundly changed the Renaissance conception of nature as the visible expression of divine order. The world is now conceived as a *machina mundi* a vast, mechanistic system (Bade, 2017) and nature becomes accessible to scientific investigation and instrumental control. The Aristotelian view that nature or natural beings possess an intrinsic purpose is replaced by a mechanistic perspective that identifies attributes such as size, duration, shape, movement, and penetrability as "primary properties" inherent in every object. In contrast, warmth, colour, tone, odour, and taste are regarded as "secondary properties", reflecting how an object appears to or affects an observer. Hypotheses and experiments are applied to isolate primary properties and discover laws that connect them to secondary properties. Gradually, experimentation and new testing tools foster the development of a technology-based industry, culminating in the emergence of power-driven machinery for manufacturing.

While the Copernican Revolution deprived humanity of its privileged position at the center of the cosmos, it simultaneously fostered the belief in scientific rationality. Guided by observation and systematic inquiry, the emerging sciences laid the intellectual foundations for the Enlightenment (1685–1815). On a practical level, the mechanistic worldview inspired new models for instrumentalisation and domination of both inner and outer nature most notably in the development of large-scale, machine-powered industries that subjected human *physis* to the disciplined routines of factory labour (Clark, 1994). The historical reconfiguration of bodies and industry labour is reflected in Foucault's analysis of disciplinary power: "*Discipline increases the forces of the body (in economic terms of utility) and diminishes these same forces (in political terms of obedience). In short, it dissociates power from the body*" (Foucault, 1977, 138).

Beginning in the 18th century, the Industrial Revolution ushered in an era of population growth, rapid urbanisation, and intensifying environmental degradation; and throughout the 19th and early 20th centuries, unrestricted economic growth, mass production, and the rise of consumer culture became central tenets of Western modernity. In this context, both nature and human beings were increasingly reduced to commodified resources within the logic of industrial capitalism (Heezik et al., 2018). Trajectories of instrumental rationality are critically examined by, for example. Adorno, who cautions not against humans' inclination to laziness, but against the relentless expansion of social and economic systems that present themselves as natural and inevitable:

> *"It is not man's lapse into luxurious indolence that is to be feared, but the savage spread of the social under the mask of universal nature, the collective as a blind fury of activity. The naive supposition of an unambiguous development towards increased production is itself a piece of that bourgeois outlook".*
>
> (Adorno, 2005, 156)

Since the mid-20th century, unregulated industrial growth has noticeably exposed the environmental and health costs of unsustainable economic practices. In response to growing environmental concerns, the early 21st century has seen a shift in how human–nature relationships are interpreted. Concepts such as ecological responsibility, sustainability, and intergenerational justice have gained prominence, prompting efforts to restructure production and consumption practices. These developments have been accompanied by international cooperation, policy reforms, and technological innovation aimed at mitigating environmental harm.

A milestone example is the United Nations' 17 Sustainable Development Goals (SDGs), adopted in 2015, which offer an integrated framework connecting environmental protection with human rights, social equity, health, and economic resilience (United Nations, 2015). While some goals, such as SDG 12, Target 12.8: *"By 2030, ensure that people everywhere have the relevant information and awareness for sustainable development and lifestyles in harmony with nature"* (United Nations, 2015) suggest a biocentric orientation, the underlying ethical framework of the SDGs remains fundamentally anthropocentric. Nature is primarily understood as a habitat for human life; responsibility is framed as care for the environment insofar as it serves human well-being; and production continues to operate within a globally capitalist system, albeit one restructured to account for ecological and social concerns.

An anthropocentric orientation is further reflected through theories of justice in the SDGs focusing on the fair distribution of environmental goods and burdens among human populations, rather than on the intrinsic value of nature itself (e.g., Fox et al., 2016; Gupta et al., 2016) and through global ethics (Collste, 2016). Some approaches derive ethical obligations from nature as a cultural and experiential space (e.g., Seghezzo, 2009), or from environmental science, as seen in Brandi (2015): *"The Earth system perspective and the planetary boundaries or guardrails framework can thus be regarded as underlining the need to reduce the existing massive inequalities in today's world"* (p. 5). In contrast, biocentrically inspired interpretations of the SDGs foreground the moral standing of non-human life, e.g. by framing nature as a legitimate "third party" in ethical deliberation (Kopnina, 2016) or by emphasising emotional and communicative dimensions of the human–nature relationship. They contend that intuitive, affective, and aesthetic engagement with nature can play a vital role in cultivating ethical values and inspiring moral action.

Ethical interpretations of the Sustainable Development Goals (SDGs), however, tend to overlook the (Bio-)Dialectics of humans as being both *part of* and *apart from* nature. This perspective is equally absent in sustainable design discourses. In response, *BioDialectics* [11] is developed throughout this book as a supplementary framework for both sustainable design and environmental ethics.

The subsequent chapter examines how sustainable design and environmental ethics align, intersect, and diverge, highlighting key challenges and opportunities in theory and practice.

1.3 Framing Challenges and Opportunities

Environmental ethics encompasses a broad range of approaches, including the study of moral beliefs and attitudes (descriptive ethics), normative theories of action (normative ethics), questions about the meaning and structure of ethical language (metaethics and critical ethics), and the moral evaluation[12] of concrete ecological phenomena. Due to the challenges of linking moral philosophy with the theoretical and practical dimensions of sustainable design, this book does not promote one fixed framework. Instead, it identifies conceptual overlaps and tendencies between environmental ethics and sustainable design. These are explored through three guiding questions: How are sustainable human–nature relationships represented? What responsibilities do humans and designers in particular hold toward nature and other living beings? And what role do production and consumption play in achieving sustainable development?

These questions lead the discussion into the normative and applied ethics domains, where the clarification and evaluation of arguments are crucial, for example whether ethical claims are logically valid and justifiable. For example, grounding rights to nature solely on its existence risks committing the naturalistic fallacy, assuming that what is natural is inevitably good. Similarly, deriving moral obligations from empirical facts alone represents the *is-ought* fallacy, which conflates ontological and normative domains (Bruening, 1971). A continuous evaluation of the justifications underlying environmental ethics and sustainable design concepts is undertaken in Chapters 2 and 3 to ensure their theoretical credibility.

Knowledge generation in design differs essentially from that in philosophy, rendering the integration of environmental ethics and sustainable design a complex and multifaceted endeavour. Unlike philosophy, design has no unified epistemic structure; it draws from diverse modes of thinking and reasoning. As Baskerville et al. (2015) argue, design research faces the dual challenge of solving real-world problems and producing new knowledge. This knowledge may be both highly context-specific and abstract, and it emerges through iterative processes of application and reflection. Furthermore, design is necessarily oriented toward creating solutions that enhance the environments in which they are applied (Brocke et al., 2020). This practical orientation carries epistemological and normative implications. Although the *is-ought* principle is traditionally considered a fallacy in formal logics, it becomes functionally valid in design, in assuming that improvement is both possible and desirable.

Normatively, design often oscillates between applied and theoretical concepts, and interpretations of what should be "improved" vary significantly. In sustainable design, most concepts situate improvement within the triple bottom line of ecological, economic, and social sustainability. However, newer theories increasingly explore phenomenological, post-human[13], participatory, or co-design perspectives and frequently criticise conventional approaches for their implicit alignment with capitalist production and consumption ideologies. Some scholars (Wijesooriya et al., 2021, Zylstra et al., 2014) emphasise the experiential dimension of the human–nature relationship, arguing that moral obligations arise from aesthetic or emotional encounters with the natural world. This experience-based argument is also present in environmental ethics. For example, Clowney (2013) argues that emotional connectedness to nature fosters empathetic responses and may serve as a foundation for morally motivated action. Advocating for biophilia[14] as an environmental virtue within an eudaimonistic[15] framework, Clowney's position is echoed in biophilic design, which proposes that nature–human connectedness leads to happier lives and that ethical sustainable design should stimulate this connection (Kellert, 2016).

However, grounding sustainable design in human experience raises important philosophical challenges. When ethical claims are grounded primarily in subjective aesthetic or emotional responses to nature, their general normative validity becomes questionable particularly when considered across diverse cultural and social contexts. And, while emotional resonance may play a significant role in developing ethical awareness, it is not sufficient to justify moral obligations or to motivate moral action. Moreover, the relationship between the flourishing of humans and nature is profoundly context-dependent and historically contingent. Consequently, idealistic or pre-critical approaches often overlook the philosophical and logical difficulties of deriving normative claims directly from human–nature encounters.

Clarifying SD concepts is a central task of this book and presents a significant challenge, as sustainable design is a vital field of inquiry for both academic and professional communities. Many foundational ideas are grounded in concrete case studies, lived experiences, and affective responses. This reliance on experiential and contextual knowledge renders analysis an inherently complex and multidimensional intellectual endeavour. Accordingly, this book engages with sustainable design in both theory and practice, revisiting key concepts with attention to the quality of ethical reasoning, openness to dialogue, and capacity for subjective engagement while remaining faithful to critical reflection and *"the unforced force of the better argument"* [16] (Allen, 2012).

The concept of *BioDialectics* is throughout the following chapters developed as a response to connect, expand, and enrich discourses on sustainable design. It is grounded in coherent epistemic and ethical reasoning while also incorporating affective, experiential, and designerly dimensions. Epistemically, a key component of BioDialectics involves connecting idiographic, experience-based knowledge such as personal narratives, with skill-based knowledge from professional practice and nomothetic, principle-based scientific knowledge (Araki, 2021). Practically, this integration facilitates among other things transdisciplinary collaboration[17] in response to the systemic nature of sustainability challenges (Keitsch, 2021). Ethically, BioDialectics resonates with a context-responsive approach, understood as a dynamic process informed by both rational deliberation and embodied, affective human–nature relationships, finding common ground with the notion of "living ethics" (Kinnunen et al., 2017). In the realms of design and art, BioDialectics contributes to the exploration of productive paradoxes, contradictions that stimulate critical reflection and creative potential, resonating for example with Guattari's *Three Ecologies* (2000) by engaging aesthetic, social, and environmental dimensions as interwoven sites of transformation within creative practice.

By embracing contradiction as a generative force, this approach opens new pathways for reimagining human–nature relationships through design practices that are both ethically grounded and experientially rich. At the intersection of ethics and epistemology, the notion of epistemic parity (Medina, 2013) allows values such as diversity, openness, and fairness to take root within sustainable design interpretations and discourses. Diversity entails, for example, an inclusive outlook that affirms the plurality of perspectives on human–nature relationships, rather than privileging a single normative viewpoint. Openness reflects a willingness to engage constructively with divergent opinions, negotiate shared objectives, and foster collaborative contributions instead of advancing narrow self-interest. Fairness refers to recognition and integration of diverse forms of knowledge particularly those that have been historically marginalised within processes of knowledge generation, communication, and decision-making. Pragmatically, BioDialectics supports the integration of diverse ecological, cultural and transdisciplinary insights into the design process (Keitsch et al., 2014). The concept attempts to contribute to the development of sustainable design theories and practices that are methodically sound, ethically informed, and practically feasible, as will be elaborated upon in Chapter 5.1.

This book encourages designer researchers, educators, and professionals to reflect on their assumptions and values regarding human–nature relationships, professional responsibilities, and the role of products, services, and systems in building sustainable societies. Identifying and realising both personal and professional values through design is presented as a vital aspect of ethical practice. From the author's perspective, one of the most urgent challenges for sustainable design in the 21st century is to support designers, particularly novices, in becoming ethically aware, embodying their values through inclusive and emancipatory dialogue[18], and realising those values through the co-creation of sustainable products, services, and systems.

1.4 Modus Operandi

The following chapters examine sustainable design and environmental ethics through the lens of philosophical knowledge, starting with this sub-chapter which presents the methodological framework and outlines the structure of the analysis. Within this examination, ethics is understood as a form of non-monological communication acknowledging that, while individuals may reflect on moral questions independently, ethical understanding attains its validity through interaction within broader societal contexts. The inter-subjective nature of ethics is further supported by the idea that it is possible, if challenging, to understand the needs, ambitions, values, and norms of others, even if one does not share them (Habermas, 1990).

Recognising that something represents a value for someone which might lead to a norm, presupposes the possibility of role reversal allowing us to imagine the position of other participants in a discourse. This implies both being able to understand others' arguments (Habermas, 1990) and (to a certain degree) being able to understand their values, ways of thinking, experiences and lifeworld. For Habermas, the universality of a shared language represents a formal *bridge principle* that allows moving from factual premises to normative assumptions via rational reasoning. However, this book places greater emphasis on the normative and hermeneutic (interpretative) potential of understanding, than communicative rationality.

Attributing intrinsic value to nature may, for example, seem illogical from a rational standpoint, while a hermeneutic approach seeks to uncover the underlying reasons and motivations for such attributions, such as their role in enriching sustainability discourse. Hermeneutics highlights the significance of idiosyncratic and socially situated knowledge. According to Araki (2021) knowledge can be categorised into four interrelated domains: epistemic, tooled, idiosyncratic, and meta-knowledge. From this perspective, the attribution of intrinsic value to nature can be seen as an expression of idiosyncratic knowledge, that is informed by communal norms, cultural narratives, and lived human–nature relationships. According to Kolb (1981) knowledge creation happens at all levels, from scientific research to philosophical reflection, and from professional expertise to daily discoveries and actions. Moral philosophy and sustainable design may rely on different disciplinary modes of learning and pursue distinct goals in the creation of knowledge (Kolb, 1984). However, both disciplines can adopt a hermeneutic approach that emphasises interpretation and understanding. This common hermeneutic ground may foster new insights into the relationship between humans and nature. Posthumanist concepts, along with speculative and critical design ideas, currently illustrate how cross-disciplinary exchange can inspire innovative thinking. One of the central aims of this book is to encourage the recognition, acceptance, and appreciation of diverse ways of thinking and experiencing the world, both within academia and in everyday life.

A methodological attempt to reconcile the different and sometimes contrasting modes of knowledge generation in philosophy and design is to interpret both as life-worldly disciplines. Hence, alongside the analytical interpretation of EE and SD concepts, this book employs narratives as a way of hermeneutically reflecting human–nature relationships. Narratives are an important part of human life and, following Ricoeur (1994), they embody a pre-linguistic understanding while also exceeding linguistic expression. A narrative can, via its explanatory potential, contribute to novel interpretations of ideas and experiences (Ricoeur, 1984) *"We have to understand something completely contrary to a copy of some pre-existing reality and speak instead of a creative imitation"* (ibid., 45). Narratives are partly relieved from the obligation of strict logical reasoning. While they connect values, norms, and actions, their openness and improvisational character can encourage readers to move beyond disciplinary epistemes. They are employed across various fields to provoke reflection or to deliberately estrange the recipient from conventional ways of thinking.

Within philosophy, narratives frequently convey reflections and inquiry. Blumenberg's *Shipwreck with Spectator* (1996) presents for example a book-length metaphor navigating existential questions:

> *"It is true that what the spectator sees in his past insofar he has been able to become a spectator at all, to learn to love the wisdom of a standpoint withdrawn from life. But what he sees also lies before him in the future as*

> *that which inevitable proceeds from life, which is a sea full of rocky cliffs and whirlpools. He avoids these with care and caution, although he knows that the success of all effort and skill expended in making his way through only brings him nearer to the point at which his crack up becomes inevitable. He knows that this very way, he comes with every step forward closer to, indeed, he even steers towards, the greatest the total, inevitable, and incurable shipwreck: death".*
>
> (1996, 63)

The use of narratives in design serves at least two main purposes. Coercive narratives functionally connect stories to users and stakeholders, illustrating how they interact with products, services, or systems, with the goal of developing solutions that have greater user appeal (Keitsch, 2012). In contrast, emancipatory narratives aim to open up new possibilities for all stakeholders involved. They regard design not merely as the planning and development of products, services, or systems, but as a practice that fosters empathy, critical self-reflection, and transformation. This dual application of narratives in design aligns with a performative epistemic paradigm, wherein meaning and impact are understood to emerge through action and interaction, rather than through static representations or objects (Østern et al., 2023).

This book comprises an exploratory literature analysis (Stebbins, 2011) of sustainable design concepts in Chapters 2 and 3 and presents narratives in Chapter 4 to illustrate ideas in real-world contexts. An exploratory analysis does not require preformulated hypotheses but needs interpretative adaptability when working with the material to provide novel insights e.g., on relations between SD and EE. The analysis focuses on three main themes found in both environmental ethics and sustainable design: views of human–nature relationships, interpretations of human and professional responsibilities, and products, services, and systems perceptions in relation to sustainable development.

The narratives presented in Chapter 4 are constructed by the author to explore ideas, beliefs, values, and actions related to the thematic concerns outlined above. These narratives are not direct transcripts or representations of participants' voices; rather, they are interpretive accounts, generated and shaped through the author's synthesis of in-depth interviews, participant observation, document analysis, and personal experiential engagement with academic sustainable design programmes and intentional sustainable communities. While respondent commentaries are incorporated where possible to lend contextual grounding, the narratives primarily reflect the author's analytical framing and narrative construction and should not be read as verbatim accounts or definitive representations of the participants' own perspectives.

Chapters 5 and 6 return to conceptual analysis. They explore values such as justice, equity, responsibility, interspecies agency, and embodiment through the lens of BioDialectics, and examine the potential benefits and challenges of integrating posthuman thinking into sustainable design research and education.

The concepts and examples from practice presented in this book illustrate the multiplicity of reflections on ethics, sustainability, and design. This book does not, therefore, promote a "best" ethical approach, but pursues arguments and the negotiation of values, duties, and actions in a rhizomatic manner (Deleuze and Guattari, 1998). Rhizomatic thinking responds to both the dynamic development of EE and SD concepts over time and their interwovenness. Some concepts "fail" or reveal unwanted consequences, while others that seemed alien or insignificant at first display sustainability-relevant impact (Keitsch, 2021). Conclusively, the rhizomatic structure of this book with its various entries into theory and practice is a tribute to the heterogeneities of environmental philosophy, sustainable design, and real-life narratives, advocating their co-existence as a necessary condition for generating future knowledge for sustainable societies.

Notes

1 Clarifying key ethical terms is essential for an informed discourse in environmental ethics and sustainable design, despite potential conceptual overlaps among these terms: *Values* are fundamental ideas considered as inherently significant; they form the basis of ethical reasoning. *Principles* are derived from values and serve as broad moral guidelines that inform normative judgments. *Norms* arise from principles as socially or institutionally accepted rules that define expected actions. *Maxims* are moral rules that individuals use when making ethical decisions. *Actions* represent the concrete outcomes through which ethical commitments are either upheld or violated (see e.g. Copp, 2002). Within an anthropocentric setting, human flourishing would for example represent a core value, forming the basis of the principle of environment protection to support health. Norms are, for example, urban planning policies that promote green spaces, maxims are to support tree planting individually because it makes the city cleaner and more liveable, actions that municipalities initiate tree planting to enhance residents well-being.

2 *"This is an era of specialists, each of whom sees his own problem and is unaware of or intolerant of the larger frame into which it fits. It is also an era dominated by industry, in which the right to make a dollar at whatever cost is seldom challenged. When the public protests, confronted with some obvious evidence of damaging results of pesticide applications, it is fed little tranquilizing pills of half-truth. We urgently need an end to these false assurances, to the sugar coating of unpalatable facts"*(Carson, *Silent Spring*, 1962, 189).

3 Biocentrism regards nature as valuable in itself, while anthropocentrism regards it as valuable insofar as it serves human interests. However, both perspectives are grounded in human experience, interpretation, and engagement with the natural world, through which such values are recognized and articulated.

4 "Epistemic" means how knowledge is acquired, "epistemological" means the study of knowledge concepts and aspects. (Stanford Encyclopedia of Philosophy, 2005).

5 Intrinsic from Latin: "intrinsecus" – internal, inside. Descriptive characteristics of physical qualities. Different to "inherent" from Latin: "inhaerere" – be closely connected with. Prescriptive characteristics such e.g., dignity, reason, beauty, etc. (Vetlesen, 1995).

6 Aristotle (*Physics*, Book II) uses the term natura naturalis to refer to what occurs through nature, not through human intervention. Spinoza (1677, *Ethics*, Part I) distinguishes between *natura naturans* – nature as an active, self-causing force, and *natura naturata* – phenomena. While natura naturans is the cause, natura naturata is the effect.

7 Like many ancient Greek terms, *eudaimonia* is a compound word with many connotations: *eu* means "good" or "well"; *daimōn* refers to a "spirit", "divine power" or "guiding force" and the suffix *–ia* indicates a state or condition. Etymological Online Dictionary (2025).

8 *"Telos"* refers to final cause of natural beings but also of human actions. For example, the telos of a plant is to grow and blossom, the telos of a human is to live a virtuous life, the telos of society to support citizens in living a such a life, etc.

9 E.g., Colossians 1:16 *"for in him all things were created, in heaven and on earth, visible and invisible, whether thrones or dominions or principalities or authorities — all things were created through him and for him".* Col 1:16 Cross References (26 Verses) (openbible.info).

10 Existentialist turn: Augustine was the first to develop a concept of the "subject" that is, of individual existence, interpretation, and subjectivity. This marked a foundational shift towards understanding the self as a being shaped by inner experience, moral struggle, and personal reflection. See e.g., Taylor, C. (1989) *Sources of the Self: The Making of the Modern Identity*, Cambridge, Mass.: Harvard University Press.

11 The concept of BioDialectics is introduced by the author of this book as a theoretical framework to explore the dynamic, reciprocal, paradoxical and contradictory relationships between human beings and the natural world. Whereas earlier uses of the term most notably in Esposito's *Bios: Biopolitics and Philosophy* (2008) primarily emphasise the tensions among human life, nature, and political power, this work offers a reconceptualisation of BioDialectics. It draws on diverse intellectual traditions, including environmental philosophy, critical theory, and new materialism, to broaden and deepen its analytical scope. A more detailed elaboration of BioDialectics is presented in the following chapters.

12 Gensler classifies ethics roughly in four categories: Descriptive ethics studies moral beliefs and attitudes, normative ethics studies ethical action, critical ethics (metaethics) examines presuppositions of moral ideas e.g. as moral semantics, what people mean by words like "good, bad, right and wrong", and applied ethics discusses particular societal phenomena from a moral standpoint (Gensler, 2017).

13 Inspired by e.g.: Haraway, D. (2015), "Anthropocene, Capitalocene, Plantationocene, Chthulucene: Making Kin". *Environmental Humanities*, vol. 6.

14 The term biophilia was coined by E. Wilson (1984). Wilson attributes to biophilia an *"innate emotional affiliation of human beings to other living organisms"* (1993, 31).

15 For different environmental ethics concepts see Chapter 2.

16 This principle for engaging in open, rational discourse grounded in deliberation rather than coercion, aligns with Habermas's (1990) framework of differentiated knowledge claims, and serves as a transcendental-pragmatic compass for interpreting sustainable design and environmental ethics throughout this book. Habermas distinguishes three primary types: Claims to truth, which are empirically verifiable and pertain to objective states of affairs in the natural world; Claims to normative rightness, which are subject to argumentative debate (pro and contra) and concern social norms and intersubjective relations, and Claims to truthfulness, which are grounded in subjective self-understanding and reflect individual experiences, intentions, and authenticity.

17 Transdisciplinarity is characterised by the collaboration between disciplinary experts and societal stakeholders, aiming to integrate diverse forms of knowledge to address complex real-world problems (Moser, 2016; Muñoz-Erickson et al., 2017; Schneider et al., 2021).

18 Discourses, where all participants can speak, question, argue, respond, and suggest freely despite their role, gender, ethnicity, etc.

References

Adorno, T. W. (2005). *Minima moralia: Reflections on a damaged life* (E. F. N. Jephcott, Trans.). Verso.

Allen, A. (2012). The unforced force of the better argument: Reason and power in Habermas' political theory. *Constellations*, 19(3), 364–381.

Aquinas, T. (1920). *Summa theologiae* (Fathers of the English Dominican Province, Trans.). New Advent. https://www.newadvent.org/summa/3001.htm#article3. Retrieved 24 March 2023.

Araki, M. (2021). Four components of expertise (SSRN Scholarly Paper No. 3806101). Social Science Research Network. https://papers.ssrn.com/sol3/papers.cfm?abstract_id=3806101. Retrieved 24 March 2023.

Aristotle. (2009). *Nicomachean ethics.* The Internet Classics Archive. http://classics.mit.edu/Aristotle/nicomachean.html. Retrieved 24 March 2023.

Aristotle. (n.d.). *Physics* (Book II, Part 8). Logos Virtual Library. https://www.logoslibrary.org/aristotle/physics/2_8.html. Retrieved 20 December 2025.

Augustine. (2005). *On Christian belief (True religion)*. Augustinian Heritage Institute. https://wesleyscholar.com/wp-content/uploads/2019/04/Augustine-On-Christian-Belief.pdf. Retrieved 24 March 2025.

Bade, D. (2017). Humanist machines: An integrationist critique of mechanical models. In A. Pablé (Ed.), *Critical humanist perspectives: The integrational turn in philosophy of language and communication* (pp. xx–xx). Routledge.

Baskerville, R. L., Kaul, M., & Storey, V. C. (2015). Genres of inquiry in design-science research. *MIS Quarterly*, 39(3), 541–564.

Blumenberg, H. (1996). *Shipwreck with spectator: Paradigm of a metaphor for existence* (S. Rendall, Trans.). MIT Press.

Brandi, C. (2015). Safeguarding the earth system as a priority for sustainable development and global ethics: The need for an earth system SDG. *Journal of Global Ethics*, 11(1), 32–36.

Brocke, J.vom, Hevner, A., & Maedche, A. (2020). Introduction to design science research. In J.vomBrocke, A. Hevner, & A. Maedche (Eds.), *Design science research: Cases.* Springer.

Bruening, W. H. (1971). Moore and "is–ought". *Ethics*, 81(2), 143–151.

Carson, R. (1962). *Silent spring*. Houghton Mifflin.

Clark, G. (1994). Factory discipline. *The Journal of Economic History*, 54(1), 128–163.

Clowney, D. (2013). *Environmental ethics: Theoretical and applied* (2nd ed.). Oxford University Press.

Collste, G. (2016). (2nd ed.). *Global ethics: An introduction* (2nd ed.). Routledge.

Copp, D. (2002). (Ed.). *The Oxford handbook of ethical theory.* Oxford University Press.

Deleuze, G., & Guattari, F. (1998). *A thousand plateaus: Capitalism and schizophrenia* (B. Massumi, Trans.). Continuum.

Ducarme, F., & Couvet, D. (2020). What does "nature" mean? *Palgrave Communications*, 6, Article 14.

Ehrlich, P. R. (1968). *The population bomb.* Ballantine Books.

Esposito, R. (2008). *Bios: Biopolitics and philosophy.* University of Minnesota Press.

Etymological Online Dictionary. (2025). Eudaemonic (definitIon). https://www.etymonline.com/word/eudaemonic. Retrieved 20 December 2025.

Fiore, R. N. (2016). Bioethics: Environmental. In H. ten Have (Ed.), *Encyclopedia of global bioethics* (pp. xx–xx). Springer.

Foucault, M. (1977). *Discipline and punish: The birth of the prison* (A. Sheridan, Trans.). Allen Lane.

Fox, O., & Stoett, P. (2016). Citizen participation in the UN sustainable development goals consultation process: Towards global democratic governance? *Global Governance*, 22, 555–574.

Fuad-Luke, A. (2009). *Design activism*. Earthscan.

Garland, K. (2000). *First things first 2000: A design manifesto*. https://www.firstthingsfirst2020.org/. Retrieved 10 May 2023.

Gensler, H. J. (2017). *Ethics: A contemporary introduction*. Routledge.

Guattari, F. (2000). *The three ecologies* (I. Pindar & P. Sutton, Trans.). Athlone Press.

Gupta, J., & Vegelin, C. (2016). Sustainable development goals and inclusive development. *International Environmental Agreements*, 16, 433–448.

Habermas, J. (1987). *The theory of communicative action* (Vol. 2: *Lifeworld and system: A critique of functionalist reason*; T. McCarthy, Trans.). Beacon Press.

Habermas, J. (1990). *Moral consciousness and communicative action* (C. Lenhardt & S. W. Nicholsen, Trans.). Polity Press.

Haraway, D. (2015). Anthropocene, Capitalocene, Plantationocene, Chthulucene: Making kin. *Environmental Humanities*, 6(1), 159–165.

Hassenzahl, M. (2003). The thing and I: Understanding the relationship between user and product. In M. A. Blythe, K. Overbeeke, A. F. Monk, & P. C. Wright (Eds.), *Funology*, Human–Computer Interaction Series, Vol. 3. Springer.

Hauskeller, M. (2005). Telos: The revival of an Aristotelian concept in present-day ethics. *Inquiry*, 48(1), 62–75.

Heezik, Y.van, & Brymer, E. (2018). Nature as a commodity: What's good for human health might not be good for ecosystem health. *Frontiers in Psychology*, 9, Article 16737.

Keitsch, M. (2006). Changing technology perceptions. *Progress in Industrial Ecology: An International Journal*, 3(1–2),41–58.

Keitsch, M. (2012). Personas: A tool for integrating the user perspective in companies' product development. *IDBM Papers*, 2. Aalto University.

Keitsch, M., & Singh, B. (2014). Cultural sustainability and the negotiation of public space: The case of Indrachowk Square, Kathmandu, Nepal. *Journal of Sustainable Development*, 7(3), 129–137.

Keitsch, M. (2021). Transdisciplinary collaboration and ethics. In M. Keitsch & W. Vermeulen (Eds.), *Transdisciplinarity for sustainability*. Routledge.

Kellert, S. R. (2016). Biophilic urbanism: The potential to transform. *Smart and Sustainable Built Environment*, 5(1), 4–8.

Kinnunen, K., Lyytimäki, J., & Ojala, A. (2017). *Living ethics: Ethics in everyday life*. Springer.

Kolb, D. A. (1981). Learning styles and disciplinary differences. In A. W. Chickering (Ed.), *The modern American college* (pp. xx–xx). Jossey-Bass.

Kolb, D. A. (1984). *Experiential learning: Experience as the source of learning and development*. Prentice-Hall.

Kopnina, H. (2016). The victims of unsustainability: A challenge to sustainable development goals. *International Journal of Sustainable Development & World Ecology*, 23(2), 113–121.

Kousis, M. (2002). Ecological marginalization in rural areas: Actors, impacts, responses. *Sociologia Ruralis*, 42(1), 86–108.

Lovelock, J. E., & Margulis, L. (1974). Atmospheric homeostasis by and for the biosphere: The Gaia hypothesis. *Tellus*, 26(1–2),2–10.
Malthus, T. R. (1798). *An essay on the principle of population*. http://www.esp.org/books/malthus/population/malthus.pdf Retrieved 24 March 2023.
Manzini, E. (2006). Design, ethics and sustainability: Guidelines for a transition phase. *Cumulus working papers*. University of Art and Design Helsinki.
Meadows, D. H., Meadows, D. L., & Randers, J. (1972). *The limits to growth: A report for the Club of Rome"s project on the predicament of mankind*. Universe Books.
Medina, J. (2013). *The epistemology of resistance: Gender and racial oppression, epistemic injustice, and resistant imaginations*. Oxford University Press.
Moser, S. C. (2016). Can science on transformation transform science? Lessons from co-design. *Current Opinion in Environmental Sustainability*, 20, 106–115.
Muñoz-Erickson, T. A., Miller, C. A., & Miller, T. R. (2017). How cities think: Knowledge co-production for urban sustainability and resilience. *Forests*, 8(6), Article 203.
Norman, D. A. (2004). *Emotional design: Why we love (or hate) everyday things*. Basic Books.
Østern, T. P., Jusslin, S., Nødtvedt Knudsen, K., Maapalo, P., & Bjørkøy, I. (2023). A performative paradigm for post-qualitative inquiry. *Qualitative Research*, 23(2), 272–289.
Papanek, V. (1991). *Design for the real world: Human ecology and social change* (2nd ed.). Thames & Hudson.
Plato. (n.d.). *Critias*. The Internet Classics Archive. http://classics.mit.edu/Plato/critias.html. Retrieved 24 March 2023.
Ricoeur, P. (1984). *Time and narrative* (Vol. 1). University of Chicago Press.
Ricoeur, P. (1994). *Oneself as another*. University of Chicago Press.
Rolston, H., III. (2006). Intrinsic values in nature. In Á. Sigurjónsdóttir & P. O. Jónsson (Eds.), *Art, ethics and environment: A free inquiry into the vulgarly received notion of nature* (pp. xx–xx). Cambridge Scholars Press.
Sanders, E. B.-N. (2002). From user-centered to participatory design approaches. In J. Frascara (Ed.), *Design and the social sciences* (1st ed., pp. 1–8). CRC Press.
Schneider, F., Tribaldos, T., Adler, C., Biggs, R., de Brémond, A., Buser, T., Krug, C., Loutre, M., Moore, S., Norström, A., Paulavets, K., Urbach, D., Spehn, E., Wülser, G., & Zondervan, R. (2021). Co-production of knowledge and sustainability transformations: A strategic compass for global research networks. *Current Opinion in Environmental Sustainability*, 49, 127–142.
Seghezzo, L. (2009). The five dimensions of sustainability. *Environmental Politics*, 18(4), 539–556.
Seneca. (n.d.). *Moral letters to Lucilius* (Letter 91, Section 9). Wikisource. https://en.wikisource.org/wiki/Moral_letters_to_Lucilius/Letter_91.Retrieved 23 March 2023.
Spinoza, B. (1677). *Ethics* (Part 1). Project Gutenberg. https://www.gutenberg.org/ebooks/919. Retrieved 23 March 2023.
Stanford Encyclopedia of Philosophy (2005). Epistemology. https://plato.stanford.edu/entries/epistemology. Retrieved 30 December 2025.
Stebbins, R. A. (2011). *Exploratory research in the social sciences*. Sage.
Swann, C. (2002). Action research and the practice of design. *Design Issues*, 18(1), 49–61.
Taylor, C. (1989). *Sources of the self: The making of the modern identity*. Harvard University Press.

United Nations. (2015). *Sustainable development goals.* https://www.undp.org/sustainable-development-goals. Retrieved 23 March 2023.

Vetlesen, A. J. (1995). Value in nature: Intrinsic or inherent. In *Rethinking deep ecology.* Centre for Development and the Environment Studies, University of Oslo.

White, L., Jr. (1967). The historical roots of our ecologic crisis. *Science*, 155(3767), 1203–1207.

Wijesooriya, N., & Brambilla, A. (2021). Bridging biophilic design and environmentally sustainable design: A critical review. *Journal of Cleaner Production*, 283, Article 124603.

Wilson, E. O. (1984). *Biophilia: The human bond with other species.* Harvard University Press.

Wilson, E. O. (1993). Biophilia and the conservation ethic. In S. R. Kellert & E. O. Wilson (Eds.), *The biophilia hypothesis* (pp. xx–xx). Shearwater Books.

World Commission on Environment and Development. (1987). *Our common future.* Oxford University Press.

Zylstra, M. J., Knight, A. T., Esler, K. J., & Le Grange, L. (2014). Connectedness as a core conservation concern: An interdisciplinary review of theory and a call for practice. *Springer Science Reviews*, 2(1), 119–143.

2 Environmental Ethics and Initial Sustainable Design Concepts

2.1 Environmental Ethics and Moral Philosophy

The preceding chapter analysed the core similarities and differences between biocentric and anthropocentric viewpoints, which, within moral philosophy, can be broadly classified into consequentialist and deontological positions. The following subchapters examine these classifications in terms of their underlying aims, moral principles (and specific maxims) and their interpretation within the context of environmental ethics.

Consequentialist ethics such as teleological ethics, utilitarianism, eudaimonistic ethics, or ethics of responsibility connect their principles to the result of an action: act in such a way that the specified (moral) goal is achieved. Deontological ethics such as ethics of duty or ethics of justice connect their principles to the motivation of an action independent of the consequences: act to realise the good for its own sake.

Moral principles of Utilitarianism are that the consequences of an action are measured by whether they serve a common good, which implies a decision on what this good comprises. The most moral action produces the most "good" for most partners involved. The "good" of an action can for example be to gain pleasure and avoid pain. The aim of "greatest pleasure for the greatest possible number" seems nonetheless problematic, among other considerations, because it is difficult to commensurate values. From an anthropocentric position, if the consumption of meat pleases, for example, many consumers, while the alternative of organic farming pleases some farmers, utilitarianism would have to make these values comparable. Another challenge is favouring the pleasure of a majority regardless of the misery of a minority.

What makes utilitarianism valuable for EE is its extension to the non-human world such as involving animals as sentient beings in the debate, as e.g., Bentham (1789) points in an early discussion on animal rights: *"The question is not, Can they reason?, nor Can they talk? but, Can they suffer"*. (ibid., 144). Newer authors, such as Singer (1990), argue that animals have "interests" that originate in their capacity for feelings. This utilitarian position sees equality as a the main good, and because both humans and animals have interests, they have the same rights with the following maxim to balance these. Linking interests and rights,

DOI: 10.4324/9781003384090-2

Singer pleads for a pathocentric ethic (gk pathos – pain, suffering): *"pain is pain, and the importance of preventing unnecessary pain and suffering does not diminish because the being that suffers it is not a member of our species"* (ibid., 220). Singer's ethics, however, faces a challenge regarding the concept of interest: although animals may have an interest in avoiding pain, other ethicists question whether this interest can be related to comprehension and choice. Frey (1989) highlights the argument by drawing on distinct semantic interpretations of the term "interest": *"Y is in X's interest"* and *"X has an interest in Y"*. The first assertion, e.g. *"a forested environment is in the interest of the moose"* seems reasonable based on zoological knowledge about the moose; however, the second assertion is linked propositionally to perceptions and choices. It is difficult to claim that moose *have* an interest in a forested environment, if the term "interest" presupposes that moose must be able to comprehend different environments, make a choice that a wooded environment is preferable and articulate this choice.

Frey's critique highlights a key dilemma in several environmental ethics and sustainable design concepts: the attribution of human features, values, motivations, and behaviours to non-human beings such as animals, artefacts, or natural phenomena.

This tendency is also referred to as anthropomorphism. From a phenomenological perspective, anthropomorphism involves perceiving human-like forms or intentions in natural elements, such as seeing figures in snow-covered trees. Intentional anthropomorphism attributes emotional states or intentions to these figures (or other non-human beings), such as assuming they wish humans well or ill (Airenti, 2018). While Frey's objection and critique of anthropomorphism is logically valid, it is grounded in a narrow semantic interpretation that relies exclusively on rationalist criteria for preferences and choices. How recognising expressions of well-being and suffering in non-human beings expands the traditionally human-centred focus of environmental ethics presenting both conceptual advantages and philosophical challenges is explored in more detail in Chapters 2.2. and 5.2.

The goal of utilitarianism in moral action is to maximise pleasure and minimise pain; however, this is often viewed as inadequate as a human life purpose. Accordingly, a consequentialist branch, eudaimonism, declares a "fulfilling life" to be the highest good that can be achieved through moral action. Eudaimonism associates moral action with virtues such as generosity, responsibility, and justice that can be learned and cultivated as part of an "art of living", as MacIntyre (1981) points out.

> *"Aristotle has cogent arguments against identifying that good with money, with honour or with pleasure. He gives to it the name of eudaimonia – as so often there is a difficulty in translation: blessedness, happiness, prosperity. It is the state of being well and doing well in being well, of a man's being well-favoured himself and in relation to the divine".*
>
> (ibid., 48)

A key weakness of Eudaimonism is the lack of distinction between the individual and collective good life; not every well-being is ethically relevant and not every form of a fulfilling life is moral. In EE this can be illustrated by the "Green Revolution" starting in the 1950s, with the goal to eliminate hunger outbreaks and improve life quality in India. Using grain hybrids, chemical fertilisers, and mechanical equipment resulted in a doubling or even tripling of the grain harvest per acre, per season, over a 20–30-year period. The Green Revolution produced immense amounts of food, but this did not significantly reduce the hunger of the poor. Besides huge ecological impacts, living conditions in some cases got even worse than before (Shiva, 2016).

Several EE concepts refer to teleological arguments, inherently tied to consequentialist ethics and vice versa. A teleological argument *relates* moral action to a purpose or a goal, while consequentialism *evaluates* moral actions if this goal is reached. In Jonas' *The Imperative of Responsibility* (1996), the moral goal is directly related to the environmental crisis and hence formulated as a negative imperative: not to endanger or destroy the possibility of future human life. This goal indirectly implies care for nature and for future generations to experience it. The onset for Jonas' anthropocentric moral philosophy is not the maximisation of happiness or a fulfilling life, but the minimisation of damage. The responsibility for actions relates to the foreseeable consequences that these actions might have, and it increases parallel to the power of humans. At the same time, it is becoming increasingly difficult to assess the effects of human actions. Because of this difficulty, Jonas proposes a "heuristic of fear" (1996) as strategy for action: to refrain from technical developments until their long-range effects are known and to give priority to pessimistic forecasts over optimistic ones when choosing strategies. Jonas' basic moral maxim is that the consequences of actions must be compatible with the continued existence of human life on earth. The moral subject is responsible for everything subject to their power. Duties arising from the "principle of responsibility" involve self-imposed restrictions, not only in the direct use of power but also in its acquisition, for example, through science and research.

In terms of biocentrism, Fraser-Darling's (1969) concept is grounded in teleological reasoning as well. He claims moral respect for all living beings because of the "purposefulness" of individuals and the biotic community. Purposefulness means here: the striving of living beings to realise their vital opportunities. The main value of life is intrinsic, granted by the existence of the individuals, ecosystems, or the entire nature. All beings have the same rights. Humans are part of a biotic community and as rational and moral beings, they are assigned the duty to protect other beings. Or as Taylor (1981) puts it: *"We grasp the particularity of the organism as a teleological center of life, striving to preserve itself and to realise its own good in its own unique way"* (ibid., 206).

Not achieving one's goal means failure in consequentialism. Deontic ethics on the other hand, claim that actions, despite producing a possible negative outcome, have a positive ethical value that resides in one's attitude, motivation, and will, for instance to improve the living conditions of other people or living beings, thereby contrasting utilitarianism which only values the consequences of an action. Relevant for the EE discussion are deontic ethics of justice and ethics of duty, two positions that anchor moral action in either universal rights, or in transcendental and universal moral responsibilities. Ethics of justice refers to the principle of equality. Equality relates to a definition of what is "just", here the same opportunities and access for all, for example as fair distribution of resources. Moral maxims are the protection of rights and of justice-related obligations.

Rawls (1971) develops an anthropocentric contractual model within an institutional structure of civil rights and duties. His starting point is the mutual recognition of individuals as moral agents (the moral reciprocity of the contractual partners) with the following two principles:

> *"(1) Each person is to have an equal right to the most extensive total system of equal basic liberties compatible with a similar system of liberty for all. (2) Social and economic inequalities are to be arranged so that they are both: (a) to the greatest benefit of the least advantaged, consistent with the just savings principle, and (b) attached to offices and positions open to all under conditions of fair equality of opportunity".*
>
> (ibid., 266)

The universalisation criterion for morally right action comprises a hypothetical procedure in which the participants do not know in which individual situation they find themselves and can make decisions only based on general human interests. The model is intended to ensure a starting position that enables decisions which are not influenced by subjective preferences: people have to share a common resource not knowing how much they will get individually. Regan (1986) extends the ethics of justice biocentrically to all sentient beings, ascribing inherent value to animals and hence the right to be treated with respect. *"The basic commonality is simply that each of us is the feeling subject of a life"* (ibid., p. 42). Regan grounds inherent value in individuals' subjective experiences and feelings and this allows him to assert basic moral rights to animals, that are not based on reasoning or voluntary actions.

The metaethical debate on which life has value and rights is dynamic and ongoing without definite answers (Omotosho, 2019). In its anthropocentric version, deontics of duty refers to humans as autonomous, free beings, who must take responsibility for their actions. This implies that human actions are not deterministic or heteronomously controlled. In Kant's ethics of duty, autonomy manifests in the will to act in line with the moral law, and to choose what is morally right because of practical, rational reasoning, which is synonymous with self-legislation. Personal self-interest must take the second place in favour of the moral law. Kant's moral law is denoted in the categorical imperative (Kant, 1993): *"Act only according to that maxim whereby you can at the same time will that it should become a universal law"* (ibid., XX)

The categorical imperative has universal validity for all rational beings, and every action must be examined by its universalisability (ibid., 30).

In EE, Spaemann (2007) holds a biocentric position on ethics of duty, that might be relevant for SDE. Facing the environmental crisis, he rejects consequentialist, especially utilitarian ethics, claiming that these only allow to sanctify bad means with the good result. His onset is ontological, humans have an innate knowledge of what is right or wrong. Life is the source of all being, and natural beings and humans share a similar ground through their existence as species (Spaemann, 2012). Spaemann proposes a self-evident telos (purpose) of nature, present in all existing, contingent beings. He believes that the telos of species has constantly been neglected by humans from the Enlightenment to the Industrial Revolution (see Chapter 1.2.). Its rediscovery will however evoke respect for nature's intrinsic values, which culminates in Spaemann's "non-functional ethic of triple reverence". *"for what is above us, what is like us and what is below us"* (translation M.K.), 1980, 198). Spaemann assumes that understanding nature's purposefulness in line with human purposefulness as a species constitutively results in responsibility towards the non-human environment. A problem with this and other biocentric claims is that insight of nature's intrinsic value does not necessarily lead to ethical action since understanding does not constitute moral implication. Comparable to Spaemann's position, Rolston's deontic biocentrism claims a reconciliation of biology and philosophy as an onset for environmental ethics, which should be non-anthropocentric yet biologically objective (Rolston 1991). His holistic nature ethics recognises an intrinsic value of nature and of all beings as parts of a natural system and hence worthy of protection. This includes inanimate matter since everything is interdependent and provides a common basis for life. Adopting a systems perspective is necessary to be able to understand individual processes, structures and individual beings conceiving them as interconnected mutually dependent. Rolston concludes that nature generates values through evolutionary processes and moral duties result from those.

Deducing ethical values from evolutionary processes is problematic in at least two respects, as Krebs (1999) points out. First, the systemic value of nature as a whole does not necessarily imply intrinsic value in individual natural phenomena or beings. Second, values that emerge through evolution can be ethically troubling from a human perspective. Viruses, for instance, are evolutionarily successful but pose serious threats to human well-being. This challenges the compatibility between biological and ethical value systems.

Even if biocentric deontics are not logically compelling, their metaethical perspective is relevant for sustainable design particularly in interpreting nature as something that exists independently of human purposes and interests. The following chapter introduces the concept of BioDialectics in relation to the biocentric notion of intrinsic value.

2.2 Prologue to BioDialectics

Western industrialisation led to a sharp rise in resource use and environmental pollution, which in turn sped efforts to develop new, environmentally sustainable ideas and technologies. Yet, science that aims to mitigate environmental harm often becomes trapped in a vicious cycle when innovations like eco-efficient technologies are driven by profit motives or trigger rebound effects, eventually compromising their intended environmental and social benefits.

Environmental challenges also gave rise to the development of environmental ethics, a branch of social ethics that seeks to re-evaluate human–nature relations. From an anthropocentric perspective, nature protection is often justified not only on ecological grounds but also through ideas of human dignity and equality. These ideas are, for example, central to the Sustainable Development Goals (United Nations General Assembly, 2015): *"We are determined to end poverty and hunger, in all their forms and dimensions, and to ensure that all human beings can fulfil their potential in dignity and equality and in a healthy environment"* (ibid., 2).

Anthropocentrism is criticised for viewing nature merely as an instrumental, rational, or recreational means. In contrast, biocentrism faces the challenge that treating nature as a paradigm and model for all life overlooks ethical understanding and moral decision-making as uniquely human activities. These activities are only possible through a separation from nature and are not intrinsic to phylogenetic processes. If ethical decisions were determined by nature for instance, by instincts then acting in accordance with nature would not present a dilemma. While the epistemic position of EE is necessarily anthropocentric, common ground exists between biocentrism and anthropocentrism in their shared recognition that the human–nature relationship comprises more than instrumental values, that humans alone bear ethical responsibility, and that meaningful encounters also with non-human beings are possible through experience and communication.

These assumptions constitute the ethical ground for the concept of BioDialectics, which explores the dynamic, reciprocal, and often contradictory relationships between humans and nature, framing them not as separate entities but as interconnected and co-evolving systems. BioDialectics aims to connect ideas of environmental philosophy, critical theory, biocentrism, and anthropocentrism to offer a more integrated understanding of human–nature relations.

Logically, all values assigned to nature are ascribed by humans, even "intrinsic" value, which does not stem from ecology or biology but from society and culture. However, this does not diminish its importance for environmental ethics. The heuristic notion of an intrinsic value of nature in BioDialectics relates to ideas of "critical anthropomorphism",[1] "autonomy", "non-instrumentality", and "non-identity" (Keitsch, 2002a). In terms of autonomy[2], Adorno's (2001) analogy of outer (phenomenological) and inner (bodily) nature and its mutual manipulation highlights similarities here negatively connoted as a loss. Adorno remarks: *"What tolerates nothing that would not be like itself, thwarts the reconciliation for which it mistakes itself. The act of violence of making something the same reproduces the contradiction which it stamps out"* (ibid., 87).

A contemporary example of human instrumentalisation is the attention economy (Franck, 2018; Marshall, 2021), in which individuals commodify their bodies, experiences, and life practices for economic profit:

> *"This process of commodification is a form of alienation, one of the most destructive elements of capitalism as it reifies those cultural processes that define human relations and activities and allows those things/processes to be converted into systems of exchange".*
>
> (Marshall, 2021, 168, 169)

The attention economy shrinks experiences of autonomy, as individual identities and social interactions are moulded into commodities shaped by market demand rather than by self-expression or authentic relationships, and nature becomes a stage for human exhibitions.

Recognising the parallels between the subjugation of inner and outer nature and the resulting loss of autonomy can foster ethical awareness, and facilitate alternative, non-instrumental interpretations of inner and outer nature. Awareness and new interpretations of nature are correlational rather than causal, yet they recognise both inner and outer nature as possessing intrinsic value, rather than merely serving as means to a (moral) end (Vetlesen, 2015).

Perceptions of "wholeness" or "otherness" in biocentric thought can be understood as interpretive responses to nature's inaccessibility and ephemerality, reflecting its non-identity: *"Natural beauty is the trace of the nonidentical in things under the spell of universal identity"* (Adorno, 1997, 73).

In an exceedingly technologised lifeworld, non-identity is as a reminder that inner and outer nature are not entirely controllable or designable, which in turn can contribute to an ethical regulative (not constitutive) towards nature: *"The reconciled condition would not annex the alien by means of a philosophical imperialism, but would find its happiness in the fact that the latter remains what is distant and divergent in the given nearness, as far beyond the heterogenous as what is its own"* (Adorno, 2001, 113). The goal of ethics:

> *"would not be to find the way back to the origin, to the phantasm of a good nature, but rather the origin would devolve to the goal, would constitute itself out of the latter. No origin except in the life of the ephemeral".*
>
> (ibid., 94)

Foucault's idea of heterotopia can be interpreted in a similar ductus exploring values of natural heterotopian spaces (Bigell, 2007). The relevance of Adorno's and Foucault's ideas to sustainable design and BioDialectics is examined in greater depth in Chapters 3 and 5.

The epistemic and normative features of BioDialectics encourage viewing human experiences of nature as complementary to rational reasoning. Intrinsic value can then relate, among other things, to experiencing the "richness" of beings and environments (Miller, 1982), or to encountering nature symbolically as a space for unmediated joy, freedom, awe, sublimity, and non-identity (Nelson, 2011; Bethelmy et al., 2019; Drenthen, 2017; Bassey, 2020).

Nature experience serves as a *regulative*, not a *corrective*, complement to rational argumentation. Ethically, experiencing nature can transcend material interpretations for example, through feeling connected with other living beings, which can evoke empathy and compassion. While logical arguments address reasoning, and normative arguments reflect on right and wrong, experience engages emotion. All three can influence ethical attitudes and motivate moral action.

Another normative feature of BioDialectics emphasises the socio-cultural and phenomenological value of natural surroundings in a lived, bodily, and everyday sense. Nature as a place of "home" and "belonging" speaks to the subjective value of a desirable environment and can inspire care for specific places. The following three perspectives represent central ideas of BioDialects.

First, the intrinsic value of nature, while epistemically rooted in anthropocentric thought, can function as a heuristic regulative principle that counters the instrumental domination of nature. Second, the experience of nature is inherently aesthetic in the original sense of *aisthesis*, as it is rooted in bodily perception and sensory engagement. Encounters with nature evoke various emotions such as joy, awe, fear, love, and compassion which are genuinely felt and often emerge spontaneously. Nature experiences cannot be deliberately designed or fully controlled. The autonomy of nature resists complete human manipulation.

Third, reflecting on and communicating experiences of nature may align with ethical values such as respect, autonomy, non-identity, and non-instrumentality. These nature discourses can serve as a complement to rationalist concepts in environmental ethics. While nature experiences may foster ethical awareness, decision-making, and action, they do not necessarily do so (Krebs, 1999). It is crucial to recognise that natural phenomena may evoke ethical reflection, but do not warrant it. There is no causal link between experiencing nature and the emergence of ethical attitudes or actions. On the contrary, assuming such a link undermines the concept of nature's non-identity, subordinating contingency to predetermined outcomes. As Früchtl (1996) notes:

> *"One and the same object allows for completely different moods. People have different biographies, and where one feels reminded of happy childhood days in a garden where bright red tulips are blooming and robins are singing, another must remember how in such a garden his aunt's throat was once cut before his eyes".*
>
> (1996, 406, Translation M.K.)[3]

The BioDialectics concept developed in this book offers an approach to bridging rational, socio-political, and lived-world discourses on human–nature relationships, individual and professional responsibilities, and sustainable lifestyles. Specific features of BioDialectics that are relevant to sustainable design are examined in Chapters 3 and 4 in relation to recent SD concepts and practices of sustainable education and living. The application of BioDialectics to sustainable design is discussed in Chapter 5.1 of this volume.

2.3 Early Concepts of Sustainable Design

The following chapter sequentially presents initial SD concepts, beginning with Papanek's philosophy.[4] These ideas were formulated with the advent of the environmental crisis, paradigmatically articulated as a growing concern for human–nature relationships, designers' responsibilities, and the environmental impact of products through resource exploitation, consumerism, and degradation. During this period, the foundational concepts of sustainable design began to take shape. Papanek offered one of the earliest maxims for sustainable design:

> *"The easiest way to save resources and energy and to reduce waste is to use less. This means consuming less, buying less, being Green, making do with what we have already – establishing an imperative to rid ourselves of all the unnecessary gadgets and applications that so hideously clutter our life".*
> (2021, Frontside of "The Green Imperative")

In his seminal work *Design for the Real World* (Papanek, 1985), Papanek articulates a strong critique of prevailing capitalist design and marketing strategies, highlighting their detrimental impact on both human well-being and the environment. He positions design as an inherently political act and urges designers to assume ethical responsibility by creating products that genuinely serve human needs, thereby addressing deeper societal and moral imperatives. This argument, however, conveys a paternalistic undertone, implying that consumers must be guided or even nudged by designers toward the "right" choices, which has drawn criticism from various authors. Lloyd (2002), for instance, argues that distinguishing between "real" and "false" benefits is inherently ambiguous. Similarly, Heskett (1980) questions the legitimacy of designers acting as universal moral agents, warning that the notion of designers as transformative prophets of human survival and renewal may be more unsettling than the very systems they seek to challenge. As he writes *"as universal social and moral prophets of human survival and regeneration, transforming not only the environment, but the very nature of man … can appear more frightening in prospect than the political and ideological status quo it proposes to replace … The reasons is that there is little evidence to suggest that designers are any less subject to frailty and error, or any better suited to effecting such a transformation than any other group"* (Heskett, 1980, 206).

The Green Imperative (Papanek 2021) expands on the human–nature relationship and introduces a dialectic between exploitation and spirituality. To engage with the spiritual value of nature and society, Papanek proposes six normative questions for designers:

> *"A positive answer to these or similar questions does not make the design visible spiritual [sic]. But the performance of such services to our fellow humans and the planet will help us inwardly. It will nourish our soul and help it to grow. That's where spiritual values enter design".*
>
> (2021, 48)

Papanek's concept of sustainable design was pioneering in the 1970s. He was among the first to explicitly connect ethical concerns such as responsibility, environmental awareness, and social justice with design methods and practice, establishing himself as one of the most influential authors in the SD field. His approach also reflected the growing critique of design's commercialisation and the affluence of the Western world at the expense of other cultures.

In contrast, the dominant design mindset since the 1950s, particularly in the United States, was largely driven by commercial interests. This attitude is exemplified in one of the earliest modern design textbooks, *Industrial Design: A Practical Guide to Product Design and Development*, which stated: *"The job of the designer is to develop product for manufacturer which will serve man better than their predecessors and to create in the consumer the desire to possess. Its ultimate goal, of course, is sales – at a profit"* (Van Doren, 1954, xi). Papanek's anti-capitalist and anti-racist stance offered a radical counterpoint and played a foundational role in shaping the emerging social design movement. His ideas promoted justice and equality through design for example, by advocating for sustainability not just for the rich but also the developing countries.

Meanwhile, conventional design thinking became more closely aligned with engineering and academic disciplines, and by the final decades of the 20th century, industrial product design had emerged as a distinct discipline, closely aligned with engineering science. This "empirical turn"[5] and growing concerns about sustainability led to the development of Design for the Environment (DfE), one of the first engineering-specific approaches to sustainable design.

In response to evolving environmental legislation, Western manufacturing industries began adopting cleaner production strategies in the mid-1980s (Frosch et al., 1989). These strategies gradually enabled the development of low-energy products and the exploration of innovative material use. Examples of such legislation include the *Sustainable Consumption and Production and Sustainable Industrial Policy Action Plan* (European Union, 2008); the *Cleaner Production Promotion Law* (United Nations Environment Programme, 2012); and *SDG 12: Responsible Consumption and Production* (United Nations Environment Programme, 2015). These initiatives aim among other objectives to promote the development of processes, products, and services that enhance resource efficiency while simultaneously reducing environmental and health-related risks (Van Berkel, 2000).

From the early 1990s onwards, industrial designers increasingly began to address the environmental impact across the entire life cycle of a product from raw material extraction to end-of-life disposal. This shift was supported by a revised life-cycle methodology introduced in 1988, which made it possible to calculate the precise environmental impacts of products and thereby facilitate the development of more sustainable design solutions. Life Cycle Assessment (LCA) evaluates all stages of a product or service, including resource and energy use, manufacturing and packaging, transportation, consumer usage, and final disposal, reuse, or recycling. By integrating traditional design considerations such as cost, quality, manufacturing strategies, and efficiency into LCAs, Design for Environment (DfE) practitioners were increasingly able to meet both corporate and consumer demand for environmentally responsible products (Unger et al., 2011).

This development contributed to the rise of the eco-design concept (Keitsch, 2015), which emphasises principles such as reducing material use, selecting environmentally friendly resources, avoiding waste and pollution, minimising the environmental impact of packaging and distribution, and designing for recycling and disassembly (Shu-Yang et al., 2004).

With few exceptions, early Design for Environment (DfE) and eco-design approaches paid little attention to ethical dimensions, such as environmental responsibility, social justice, and transforming excessive consumption patterns. However, growing criticism of technology-driven environmental design practices, especially their potential to support "greenwashing" after the turn of the millennium led to a more reflective and critical understanding of environmental and sustainable design, as articulated by authors such as Ehrenfeld (2008): *"The key to sustainability will be a balance between devices and a modified consumption … and products and services that can transparently restore the human capability for caring and coping in all dimensions of life"* (ibid., 123, 124).

Initially focused on purely eco-technological strategies aimed at environmental improvement and conservation, the discourse gradually expanded to include reflections on the human–nature relationship. The expansion of sustainable design primarily addressed the socio-cultural dimensions of sustainability as exemplified in McLennan (2004): *"Sustainable design starts with the understanding that the purpose of our design is to create physical artefacts that benefit people"* (McLennan, 2004, 5). Earlier concepts of sustainable design can be broadly categorised into two main branches: environmentally oriented approaches, which emphasise technological solutions to ecological challenges, and socio-politically oriented approaches, which focus on society, stakeholder engagement, and user inclusion. The latter builds on early 20th-century ideals that regard designers as educators of users and consumers. Beginning in the early 1970s, this role expands significantly: designers are no longer viewed solely as providers of physical solutions to users' needs (e.g., Dreyfuss, 1967), but increasingly as experts capable of interpreting users' social and symbolic needs. In 1989, Krippendorff introduced a narrative turn[6] in design with his concept of "product semantics", highlighting *"a concern for the sense artefacts make to users"* (Krippendorff, 1989, 10). Krippendorff's ideas imply a significant advancement of interpreting users' needs, and the role of designers: not only as interpreters but also as communicators of social and symbolic meanings through products and services.

In the first two decades of the 21st century, new agendas emerged, arguing among others that sustainable design should move beyond the traditional triple bottom line of ecological, economic, and social sustainability and actively promote human well-being and political stability (Fuad-Luke, 2009). Other sustainable design approaches have incorporated elements of social practice theory to actively engage users in sustainability transitions (e.g., Shove, 2003; Patterson, 2006) or interpret nature either as a model for biocentric design (Beatley, 2011) or as a source of empirical, bio-informed solutions (Benyus, 2002).

In recent years, sustainable design discourses have become increasingly diverse, incorporating theoretical perspectives from phenomenology, constructivism, and postmodernism. These concepts foreground the role of human experience, emotion, co-creation, persuasion, and speculative or disruptive thinking in shaping sustainable futures. The following chapters examine the ethical implications of sustainable design concepts in detail, with a focus on their perspectives regarding human–nature relationships, the responsibilities of designers, and the role of products.

2.4 Nature as a Model

Relating ethics to human– and natural–environment relationships implies asking for reasons, dimensions, and tendencies of what makes this relationship problematic. Several SD concepts see nature as a model and reference frame for discussing these questions. Among the huge breadth of theories considering nature as a prime source for sustainable design, either epistemologically or normatively, or both, biocentric views are most frequent. The following section discusses two biocentric concepts: "nature as a model" referring to rational problem-solving, and "nature as a metaphor" referring to emotion and experiences as guidelines in sustainable design.

Design for the Environment (DfE), and eco-design and planning concepts usually employ a "nature as a model" view, epistemically following a "learning-from-nature" paradigm. In many concepts, "nature" equals "ecosystem", which is interpreted as a network of co-dependent and ultimately viable entities, while terms such as "system", "balance", "boundary", "homeostasis", etc., are seen as patterns and principles for sustainable design interventions, e.g. in Odum (1969): *"ecological succession involves the development of ecosystems; it has many parallels in the developmental biology of organisms, and also in the development of human society. The ecosystem, or ecological system, is considered to be a unit of biological organization made up of all of the organisms in a given area (that is, 'community') interacting with the physical environment so that a flow of energy leads to characteristic trophic structure and material cycles within the system"* (ibid., 262). Systems theory is a crucial component of the ecosystem view. In DfE and eco-design, systems are formally interpreted as units with interrelating parts that perform functions, which overcome their individual limitations. The structure of a system comprises relatively stable established pathways because of continuous interactions between the parts. The pathways act as "patterns" in relation to "functions" as "actions". Functions modify the existing structures by constituting new pathways and become established structures in time, serving as templates for new action parameters and so forth. Systems evolve by becoming more complex and more "intelligent". The most sustainable systems are those which are most complex and open. Complex open systems are relatively stable because resources are processed through and assimilated. Complex systems have more and better opportunities to react to changes (Isenmann et al., 2007). Relating to natural systems, some authors also recommend that planetary boundaries should provide the fundament for global ethics (Brandi, 2015). Pichler et al. (2017) assert: *"Based on an in-depth understanding of society–nature relations, conclusions for sustainability transformations can be drawn that go far beyond the postulated need for 'Earth System governance', that is, 'societal steering of human activities with regard to the long-term stability of geobiophysical systems'"* (ibid., 34).

Many DfE and eco-design concepts consider the interdependency of human and ecological systems on an instrumental rather than a normative level. Designers have no specific ecological or social responsibility but are problem solvers. These problems are complex because human–nature systems comprise numerous tightly interlinked elements and interactions. DfE developers are pragmatists; the goal is reached if control of the process is achieved. Control means here the improvement of products' and services' environmental performances, while the social dimensions of sustainability are considered irrelevant.

In summary, rationalist and technology-oriented "nature as a model" approaches reflect two principles in varying degrees: a division between the technical and the social, based on the belief that technology has developed independently from society (depending on its own logical and functional premises) and a view that technology is an independent variable for the evolution of society (Andersen et al., 1994). DfE and eco-design as scientific "objective" approaches face criticism from at least two sides: designing in and for societies and a capitalist system that ultimately counteracts sustainability ambitions (Madge, 1997), and paying too little attention to human-related aspects, for example in the use phase (Bhamra et al., 2011). Ehrenfeld (2000), among others, responds to such critiques by first arguing that human history cannot offer suitable metaphors for sustainable living, as it represents a succession of states largely built upon one another, but nature can provide metaphors for sustainable trajectories. Second, he states that while the primary task for today's designers and planners is to meet eco-technical principles such as low material-energy intensity and high regenerative potential, it is equally important to respond to users' and societies' needs through product and service solutions (2000).

Instead of empirically interpreting terms such as "ecosystem" and "nature", authors who employ the "nature as a metaphor" perspective tend to use these terms figuratively. Papanek (1985) presupposes, for example, a "genuine reality", opposed to a "fake" one, and nature is also presented as a teacher: *"One source that never seems to go out of style is the handbook of nature. Here, through biological and biochemical systems, many of the same problems mankind faces have been met and solved. Through analogues to nature, man's problems can be solved optimally. … To put it more simply: to study basic principles in nature and then apply these principles and processes to the needs of mankind"* (1985, 186).

Papanek (2021) argues that ecological concerns oblige designers to adopt a moral stance. Nature is associated with concepts such as *harmony, balance*, and *healing*, while Western consumerism and materialistic attitudes are critiqued. In contrast to the commercialised Western figure, Papanek presents the image of the indigenous, unspoiled native such as the Balinese people as bearers of ecological wisdom and cultural knowledge (ibid., XV). This critical perspective on the mindless consumption of goods and services, and the role designers can play in resisting it, is echoed by scholars such as Bonsiepe (2006) and Tonkinwise (2003)[7]. Similarly, Van der Ryn and Cowan (2007) advocate for a systems-oriented and phenomenologically grounded ethical approach in sustainable design and planning. *Ecological Design* states:

> *"In other words, within this clockwork solar system lies the potential for the slightest perturbation to become vastly magnified. … When we compound a system's intrinsic capacity for enormously complex behavior with our own uncertain, incomplete models, we are left with a partial understanding at best. … As nature has receded from our daily lives, it has receded from our ethics. … By making nature visible again, favoring technologies that are not hidden and that do not possess hidden consequences, our imaginations are again enfolded in nature".*
>
> (ibid., 87, 186)

In summary, "nature as a metaphor" concepts comprise a critical worldview of contemporary consumption society, sometimes infused with technology scepticism and nature romanticism, presenting, for example, nature as an ultimate teacher for humans. Designers are proclaimed to have an all-encompassing moral responsibility for their actions, based on their individual capabilities to understand what is right, while societal stakeholders such as companies and consumers tend to counteract their efforts. A generic ethical maxim resulting from the above concepts is regarding nature and society as equally relevant and interdependent in sustainable design and planning. Yet, disciplines such as ecology or biology have no normative basis, as e.g., Enzensberger (1974) points out:

> *"An ecologist researching the conditions of life in a lake has solid methodological ground to stand on; ecological arguments begin to become shaky only when the ecologist involves his own species in them. Escape into global projection is then the simplest way out. For in the case of man, the mediation between the whole and the part, between subsystem and global system, cannot be explained by the tools of biology. This mediation is social, and its explication requires an elaborated social theory and at the very least some basic assumptions about the historical process".*
>
> (p. 17)

Ecosystem concepts are grounded in positively framed perceptions of ecosystems or nature, resulting in self-referential ethics. This is problematic because alienation from nature is just as perceivable as connectedness. Moreover, the actual degradation of natural ecosystems stands in direct contradiction to ontological ecosystem models and nature-based ethics. Sustainable design ethics grounded in concepts of nature or ecosystems tend to face similar paradoxes to those discussed in Chapter 1. It seems hence more reasonable to base ethical approaches on ideas of humanhood and humaneness, as in socially sustainable design (SSD) concepts, which situate ethics within integrative and emancipatory frameworks[8]. SSD concepts range from promoting sustainable behaviour through changes in everyday practices and lifestyles, to more radical approaches focused on social innovation.

2.5 Social Sustainable Design

The value of the human–nature relationship for social and community development and cultural enrichment is the normative onset for, mostly anthropocentric, social sustainable design (SSD) concepts. Fry (2009), Thakara (2005), and Bonsiepe (Fathers, 2003) argue, for example, that the development of products (goods and services) directly impacts society and the environment, and that the designers' overall task is *"to make the world more habitable"* (2003, 53). Perceptions of designers' responsibility and agency to contribute to social change "by design" engendered various social sustainable design concepts. In the following, two branches are investigated: Design for Sustainable Behaviour (DfSB), a "psychotechnological"[9] approach that investigates user behaviour and the influence that designers can exercise on this, and an "emancipatory" approach that focuses on user inclusion, participation, community well-being, and transdisciplinary design and planning collaboration[10].

A central tenet of Design for Sustainable Behaviour is the perspective that users are not merely passive recipients of design interventions, but active agents whose choices and behaviours influence sustainable consumption. Yet, users are seen as unpredictable and inconsistent individuals, and they do not always choose what is best for them or the environment. Consequently, DfSB combines eco-design strategies with interventions to modify users' behaviour.

De Medeiros et al. (2018) mention four crucial DfSB principles: *"making it easier for people to adopt a desired behaviour; (ii) making it harder for people to perform an undesired behaviour; (iii) making people not want to perform an undesired behaviour"* (ibid., 403)

DfSB concepts adopt a user-centred approach, aiming to influence individual decision-making processes by guiding, or "nudging" users toward more sustainable patterns of consumption. An inherent dilemma that is rarely discussed in DfSB concepts lies in the users' voluntariness to make changes versus being coerced via design. Voluntariness in DfSB seems to apply only to designers, while "social engineering" implications are only sporadically discussed by some authors (Brey, 2006; Pettersen et al., 2008; Bhamra et al., 2011).

The prevailing credo within the field is that: *"an element of force (in design M.K.) to ensure the change happens"* (ibid., 433) is generally acceptable. Lockton et al. (2008) present, e.g., affordance- and constraint-based methods to coerce user behaviour to sustainable consumption, referring among others to the *"six weapons of influence (reciprocation, commitment and consistency, social proof, authority, liking, and scarcity)"* (ibid., 5). This militaristic metaphor might symbolise urgency but can also indicate the user as an enemy. Fogg's tools, based on motivation, ability, and triggers, are "friendlier", yet his "Behavior Wizard" model (2010) is not less coercive: *"a systematic approach to designing for behaviour change (that, M.K.) can empower us (the designer, M.K.) to create persuasive technologies more efficiently"* (ibid., 127).

DfSB implementations can contribute to incremental changes towards sustainable societal practice, and concepts provide good indicators and benchmarking of how stimulation can influence behaviour and habits. Yet, regarding the development and dynamics of behaviours, DfSB shows some deficiencies. At the level of individual decision-making, concepts fall short in explaining how personal experiences, insights, and reflective processes contribute to behavioural change either independently of or in concert with DfSB strategies. This limits a comprehensive understanding of behaviour change as an ongoing, value-driven process that goes beyond mere external influence.

From an ethical perspective, DfSB concepts tend to place the burden of moral decision-making on designers or planners, who are tasked with guiding user behaviour through design. They are also supposed to navigate complex ethical challenges, including the potential for unintended consequences arising from the implementation of sustainable products and services. From an epistemological perspective, many designers may lack formal training in ethical theory and reasoning, which limits their capacity to anticipate or evaluate moral implications comprehensively. From a moral standpoint, designers are not inherently legitimised to make value-based decisions on behalf of users, raising critical concerns about agency, autonomy, and accountability in persuasive and behaviour-influencing design practices.

The strong tactical focus of many DfSB concepts to manipulate users psychologically into performing "sustainable" actions further displays a surpassing technocratic ideology, appearing almost as a parody of the "leave no one behind" motto, or as Habermas (1974) puts it:

> *"the danger of an exclusively technical civilization which is devoid of the interconnection between theory and (political, M.K.) praxis can be clearly grasped; it is threatened by the splitting of its consciousness and by the spitting of human beings in two classes: social engineers and the inmates of closed institutions".*
>
> (ibid., 282)

DfSB has become increasingly nuanced in recent decades, particularly through the integration of persuasive design concepts that acknowledge the intersections between environmental concerns, production and consumption, and economic interests such as capitalist market dynamics. These approaches also address socially desirable regulations as a means of mitigating corporate manipulation mechanisms. Persuasive design concepts are examined in detail in Chapter 3.6.

Reacting to sustainable practice challenges, social sustainable design (SSD) concepts attempt to incorporate various aspects of human relations into natural surroundings as motivators for sustainable actions. While often subscribing to emancipatory design and planning discourses SSD concepts aim at giving all participants an equal opportunity to contribute to a discussion, for example, by narrating their own experiences. Individual experiences of natural surroundings are used in these discourses to supplement facts and rational arguments with the goal to build a common and transdisciplinary ground for actors (Giri, 2002; McGregor, 2015).

The emphasis on lived experience reflects how humans relate to nature and, to some extent, their desire for it to be benevolent. This may be grounded in the multifaceted ways in which humans depend on nature not only physically and materially, but also cognitively, emotionally, aesthetically, socially, and spiritually. The experience and communication of natural phenomena transcend purely material concerns, helping to counter overly instrumental or utilitarian views of the environment. For this reason, Socially Sustainable Design (SSD) aims to foster more reflective, value-driven, and emotionally grounded forms of sustainable behaviour. Experience and appreciation of place can further contribute to creating identity, a common horizon, and a socio-cultural reference frame. The term "identity" connected to natural surroundings relates in SSD to notions of "place", "home", and "belonging", and to love for and attachment to specific environments.

In 1995, the United Nations formulated commitments for social sustainability, stating among others:

> *"people are at the centre of our concerns for sustainable development and that they are entitled to a healthy and productive life in harmony with the environment … We commit ourselves to promoting social integration by fostering*

societies that are stable, safe and just and that are based on the promotion and protection of all human rights, as well as on non-discrimination, tolerance, respect for diversity, equality of opportunity, solidarity, security, and participation of all people".

(1995, Annex 1, Section 8 and Commitment 4)

"Emancipatory" SSD concepts often comprise eudaimonistic principles as basic normative goals, such as the "common good life" and "community well-being". One of the most significant design concepts is Manzini's "Sustainable everyday" philosophy (2003). Manzini's worldview is optimistic and focused on foresight, creativity, and interaction: *"Indeed, we cannot act in a forward-looking way if we are unable to imagine a state in which we could potentially live in a different and more attractive way than now"* (ibid., 13). Epistemically, Manzini adopts a social constructivist view[11]; methodologically, he combines natural sciences and engineering with social sciences concepts. Designers are seen as significant actors in moving towards sustainable societies, yet in contrast to DfSB, they have no authoritative role:

"Conceiving and proposing products, services and lifestyles, designers play an important role and consequently have an equally important responsibility in generating social expectations in terms of wellbeing. ... Of course, designers have no means of imposing, for good or bad, their point of view on others. But they do have the tools to operate on the quality of things, and their acceptability, and therefore on the attraction of the scenarios of wellbeing they help to generate".

(Manzini, 2006, 2; see also Manzini et al., 2003, 233 ff)

Manzini presents practical SD guidelines in the form of two fundamental technical principles: low material-energy intensity and high regenerative potential. These principles originate in eco-design guidelines and are connected to personal and social wellbeing (Manzini, 2003). The social task for designers is: *"to conceive scenarios of wellbeing in which the overall quality of the context of life has to be considered, in which the physical and social common goods are regenerated and where contemplative time has its place"* (ibid., 14). The idea of creative communities where stakeholders interact locally in daily life to promote sustainable living is the most prominent feature of Manzini's concept:

"There is, in my view, a new model of organizing society and the production and consumption and whatever. When I use the words small, open, local and connected, this is my way of telling the story. ... For me, dealing with the needed sustainable changes that are mainly cultural and behavior change, the pivotal moment has been when I moved from saying 'What can I do to help people change behavior?' toward the discovery that a lot of people (even if they aren't yet so visible) had already changed, and in a

> *good way, their behaviours. And that therefore, the right question is: 'What can I do to trigger and support these new ways of thinking and doing? How can I use my design knowledge and tools to empower these grassroots social innovations?'"*
>
> (Manzini, 2011, n.p.)

Manzini's emancipatory approach is based on the idea that sustainable design should not only meet environmental challenges but simultaneously contribute to human wellbeing and socio-political stability. Ethically, his concept oscillates between Aristotelian Eudaimonism, emphasising the importance of citizens' social and political involvement and the mutual responsibility of stakeholders, and pragmatic meliorism[12]:

> *"New environmental schemes cannot be proposed in terms of unitary representations, as likenesses of closed worlds. Instead, one must find guiding ideas that constitute exemplars of quality and that are potential generators of an extensive totality of solutions that are diverse but coherently organized in certain respects. It is desirable that design culture advance a plurality of possibilities, diverse 'socio-environmental schemes'".*
>
> (Manzini, 1992, 20)

Interpretations of nature in social sustainable design (SSD) concepts remain bleak. Nature is frequently framed merely as a backdrop or surrounding system, while the emphasis predominantly lies on stakeholder inclusion in decision-making and the design process. This focus is typically presented as a means to create a "habitable" world in which humans can thrive. Both the inclusion process and the results are often portrayed as a demonstration of moral awareness and social responsibility on the part of designers. However, the enduring ethical expectations placed on professionals to embed values within products, services, and systems lack clear guidance on how designers should cultivate ethical reasoning or decision-making abilities, especially given the marginal role ethics plays in most design curricula (Keitsch et al., 2010). Some SSD concepts rely on common-sense morality, assuming it suffices for professionals to make ethical decisions on behalf of users and communities. Others adopt analytical approaches to moral dilemmas[13] or advocate for institutional support to help designers develop a personal ethical foundation (e.g. Findeli, 2001).

In terms of design, both Design for Sustainable Behaviour (DfSB) and Socially Sustainable Design (SSD) tend to move away from the individual focus initially emphasised in the design process, favouring standardised mechanisms or collective strategies instead. While these approaches can be effective in achieving broad behavioural shifts and changes in practice, they often come to regard individuals as abstract entities or representatives of a collective, rather than consistently acknowledging them as embodied agents with personal, emotional, and sensory relationships to their natural environments.

Addressing a lifeworldly and phenomenological perspective, deep ecology began to attract increasing interest among sustainable design scholars in the early 1990s, and core ideas are still present in contemporary sustainable design discourses (Kellert and Wilson, 1995; Dilnot, 2011; Brunello, 2025).

Rooted in ecocentric philosophy, deep ecology emphasises the significance of lived, everyday experiences with nature as the foundation for an ethical framework grounded in ontological principles. The following chapter introduces the core tenets of deep ecology and presents two sustainable design approaches influenced by its philosophy: bio-urbanism and Dilnot's concept of "sustainability in the deep sense" (2011). Rather than relying on discourse, formal education, or professional training to cultivate ethical reflection and action, these approaches posit an inherent human affinity for nature as a fundamental prerequisite for moral engagement manifesting, for instance, as acts of care, respect, and responsibility.

2.6 Deep Ecology and Sustainable Design

Deep ecology (DE) contains certain logical fallacies (Anker, 1999), which may explain its limited presence in disciplinary philosophy compared to its broader influence across other fields. Nevertheless, it holds substantial potential to enrich the discourse on sustainable design. Deep ecologists view nature biocentrically, possessing an intrinsic *telos* that has only been blurred, repressed, and overwhelmed by ideologies such as Christian dogmatism, Cartesian philosophy, and patriarchy. Hence, deep ecology often couples its critique of contemporary human relationships with scepticism toward the core values of modern civilization.

DE concepts reject the idea of humans as superior to other living beings; on the contrary, they are but one possibility of life's immeasurable diversity, and their preservation and further development is not of greater importance than that of other species. The acceptance of an immanent connection with other forms of life enables humans to identify themselves positively with the surrounding nature and its phenomena. Epistemically, this eco-compatibilism (Aiken, 1984) is challenging since nature must be presupposed as good; otherwise, humans would have to act unethically when being part of it. Furthermore, the hypostatisation of a context-dependent judgment "good for humans" into an objective moral claim "good for all" or "good in itself" presents a significant philosophical problem, for example, with regard to the naturalistic (is–ought) fallacy. Politically, eco-compatibilism can undermine human autonomy when ecological laws dictate morality, e.g., via the maxim that what is good for the individual is best achieved by realising the good in nature. Deep ecologists, however, tend to avoid the "human autonomy–nature determination" antinomy[14] by recurring to ontological ethics, i.e., ethical understanding is generated through existence (and its recognition), not because of transcendental reflections on values, norms, etc.

The Norwegian philosopher Arne Næss, who coined the term *deep ecology*, is one of the most prominent figures in the early deep ecology debate. He distinguishes between shallow and deep ecology

1 *"The Shallow Ecology movement: Fight against pollution and resource depletion. Central objective; the health and affluence of people in the developed countries".*
2 *"The Deep Ecology movement: Rejection of the man-in-environment image in favour of the relational, total-field image. Organism as knots in the field of intrinsic relations. An intrinsic relation between two things A and B is such that the relation belongs to the definitions or basic constitutions of A and B, so without that relation, A and B are no longer the same things. The total field model dissolves not only the man-in-environment concept, but every compact thing-in-milieu concept except when taken at a superficial or preliminary level of communication.*(Næss, 1989, 27, 28)

Næss advocates for principle-based egalitarianism, asserting that all beings have the same right to live and flourish, and are equal in value and status (Næss, 1989). In 1995, Næss and Sessions developed a platform for transitioning from shallow to deep ecology (Sessions, 1995). The platform comprises principles such as epistemically acknowledging the intrinsic value of all beings and its realiation via diverse forms. Ethically, humans have no rights to reduce this richness, except to meet vital needs.

As human actions towards nature become increasingly excessive and ecologically unsustainable, one proposed response consistent with the principles of deep ecology is the gradual reduction of the human population as a means to mitigate environmental degradation (Næss, 1989). However, demographic change alone is insufficient. Structural transformation at the policy level is equally crucial, as public policy holds the capacity to drive systemic shifts for example, by promoting the development and adoption of "soft" and contextually appropriate technologies that align more closely with ecological principles and long-term sustainability. Næss regards designing products as part of human nature but underlines that even if products can contribute to insights and provide experiences, they are mere means to develop a sustainable society and have no purpose in themselves. The goal of society is to be long-term sustainable in its industrial, economic, and political endeavours. Industry should operate in an "ecological system of economy"; technology must be soft or "appropriate". The term "advanced" technology signifies approaches that best meet requirements of ecologically sound production processes such as minimal use of energy, pollution, and unrenewable natural resources (Næss, 1989). In terms of social and individual values, appreciating the quality of life rather than an increased life standard is necessary. For Næss, this is expressed by the difference between "big" and "great". Finally, everyone who subscribes to the principles above is required to implement necessary changes (ibid., 68). The platform has discursive ethical relevance in enabling individuals and groups to find common denominators as a basis for ecological interaction.

Næss' deep ecology concept pursues three paths: first, the platform supports eco-philosophy as an interdisciplinary, deep ecological movement where scientists, activists, educators, and artists can participate. Second, he aims at a deep ecological philosophy that will develop if philosophers continue to elaborate the "platform" and its implications; and third, he sketches an *ecosophy* as a personal worldview, connected to personal ethical understanding and commitment. Ecology and eco-philosophy are based on descriptions of the interdisciplinary, scientific research of living conditions of organisms in relation to each other and to the organic and non-organic environment, to discover criteria for the self-constitution of organisms, their internal organisation, and their external relationships with the environment. Næss assigns no explicit prescriptive values to ecology; empirical knowledge provides information and support for discussing value questions. Ecology findings, under the maxim: "all substances are connected", intersect with philosophical questions. Philosophy is reflecting, for example, on human–nature relationships compared to other forms of life, or terms such as "evolution", "systems", "diversity", "complexity" and "symbiosis" to present concepts of co-existence (Keitsch, 2012).

Næss locates collaborations of ecology, philosophy, and other disciplines or sectors under the "eco-philosophy" umbrella. He then moves from epistemology to ontology and ethics by introducing the concept of *ecosophy*. The role of philosophy is here to supplement theoretical reflection with individual understanding of ethical and aesthetic nature values, which both guide practical decisions. While eco-philosophy can be studied in academic settings, ecosophy can only be developed phenomenologically through individual experience and practice. Everyone can develop their ecosophy, which does not have to be new, perfect, or original, but should be grounded in one's own considerations and decisions.

> *"We daily decide between conflicting considerations and interests. To work for a more ecologically responsible personal and societal lifestyle is thus not merely the ecologist's job. Nor the philosopher's. We should all do it together, as generalists rather than as specialists. We all have a certain amount of practice in choosing courses of action, and this presumes an ability to make comprehensive value judgments".*
>
> (Næss, 1989, 45)

For Næss, the development of an individual ethos comes from another source, namely from a certain way of phenomenologically experiencing nature, which is crucial for developing a non-instrumental attitude towards its beings. Experiencing nature's qualities as "joyful" will enable humans to act morally. The development from a "reflective-ecological" to a "reflexive-ecological" self (ibid., 119) happens via nature experience and evokes emotions and normative insight, which culminate in a "worldview" and subsequent ecological behaviour:

> *"Appropriate spontaneous experience and conscious understanding elicit quite naturally those behaviours which a cumbersome ethics tries to impose … A set of ethical codes has little motivating force if it is not felt also to be a code of natural behaviour when we are attentive and thoughtful and sensitive to what is going on around us".*
>
> (ibid., 6)

Næss ignores an implicit contradiction by assuming "worldview" as an overall, coherent experience of the environment. *De facto*, experiences of reality (and nature) are, however, contingent and fragmented. This contradiction can only be met if "worldview" is connoted with the spontaneity of an "aesthetic *telos*"[15], as for example in Gadamer's ontological interpretation of encountering "beauty", instead of presenting nature experience *per se* as ethical value and condition for moral action:

> *"we learn that however unexpected our encounter with beauty may be, it gives us an assurance that the truth does not lie far off and inaccessible to us but can be encountered in the disorder of reality with all its imperfections, evils, errors, extremes, and fateful confusions. The ontological function of the beautiful is to bridge the chasm between the ideal and the real".*
>
> (Gadamer, 1986, 15)

Næss' ontologically based virtue ethics, *ecosophy*, relies on identification with the individual "ecological self", achieved through the conscious perception of the global human threat to the environment, the insight into the interdependence of life, and ethical self-realization based on acceptance, affection, and helpfulness towards other life (Sessions, 1995). Ontological ethics can be criticised for obviating the need for a societal discourse about what is morally right and what is not. If individuals are considered as ultimate authorities, not only competent to know what is good for them but also what is right for the community (of all living beings), discussing opinions is pointless. Næss argues that everyone who has a good ecological self and works in a community contributes to the benefit of all, and suggests common therapeutic, eco-psychological healing to develop sensitivity to the environment and find beneficial ways for community interactions (Næss and Drengson, 2008). This "therapy" idea resulted in a deep ecology branch that aims to reveal the ecological unconscious, since its suppression is assumed to be the innermost root of a collective madness in industrial societies. Via psychological and spiritual methods, the latent sense of ecological interdependence should be reawakened, and the ecological self should tune in to the planet. Macy asserts:

> *"Just as lovers seek for union, we are apt, when we fall in love with our world, to fall into oneness with it as well. Hunger for this union springs from a deep knowing to which mystics of all traditions give voice".*
>
> (2007, 27)

As previously noted, DE concepts are subject to several critiques Besides the epistemological fallacy of organicism, which becomes a natural fallacy when transferred to the moral dimension, deep ecology is politically problematic. Some authors disapprove, e.g., that organicism claims a natural order via evolution, and find this assertion hierarchical and anti-democratic. The individual is subdued under an ideological, quasi-spiritual whole, and harmony is enforced instead of society solving conflicts with the help of rational and emancipatory discourses and decision-making. Zimmermann (1995) argues that this renders deep ecology susceptible to extreme environmentalism:

> *"According to deep ecologists, the longer democratic societies postpone making the difficult political decisions needed to solve environmental problems, the more drastic will be the political measures that may have to be taken later on to save remnants of humankind and the biosphere. One can imagine what such measures might entail: not only the seizure of private property, but perhaps also harassment, internment, torture, deportation, and worse, those designed to force people to comply with centrally-imposed regulations (in areas ranging from consumption to reproduction) purporting to deal with an "ecological emergency".*
>
> (1995, 209)

Socio-political critics, such as Bookchin, emphasise that deep ecology overlooks social, ethnic, and gender differences by assuming a false notion of human equality.

> *"Deep ecology, despite all its social rhetoric, has virtually no real sense that our ecological problems have their ultimate roots in society and in social problems. It preaches a gospel of a kind of 'original sin' that accurses a vague species called humanity – as though people of colour were equitable with whites, women with men, the Third World with the First, the poor with the rich, and the exploited with their exploiters".*
>
> (Bookchin, 1999, 283)

Despite these objections, deep ecology has had significant influence on a range of design theories and practices. Its core principles especially the emphasis on ecological interconnectedness and the intrinsic value of non-human life, have informed concepts such as biophilic design and bio-urbanism, which focus on enhancing human–nature relationships and align with deep ecology's quest for a fundamental transformation towards sustainable societies.

Bio-urbanism perceives cities as "living organisms" with the goal to enhance harmonious urban growth. An instrumental, systems-oriented branch of bio-urbanism promotes, e.g., a biophilic view by connecting built environments with human life-quality and community development (Salingaros, 2000). Socially oriented bio-urbanism focuses on psychological, emotional, social, and cultural qualities of (urban) nature as a source of health and well-being. The term "biophilia" connotes a "deep-seated need of humans to connect with nature" (Tran, 2017, 201). Biophilic design materialises human–nature connections by integrating natural elements in social surroundings, implementing:

> *(1) a naturalistic dimension; (2) wholeness of the site, and (3) geometric coherency. … All of these attributes reinforce social interaction by supporting the unfolding of inclusive environments that reflect the inborn affinity between human being and nature, and at the same time, are capable of supporting our neuro-physiological, psychological, and biological systems.*
>
> (Caperna, 2018, 9)

Deep ecology and biophilic concepts both draw upon ontological ethics by claiming that accepting the essential necessity of specific natural surroundings for humans induces care. Attentiveness to living beings that share the same surroundings translates in some biophilic concepts to caring for the intertwined needs of humans and nonhumans within interdependent communities (Hung, 2017). While care in deep ecology means acknowledging, accepting, and cultivating individual identification with the lifeworld and hence achieving an ethical mindset *en passant*, biophilic concepts see care as active collective responsibility:

> *"With our knowledge and a sense of responsibility for the welfare of humankind and the earth, we can create new environments that are ecologically sound, aesthetically satisfying, economically rewarding, and favorable to the continued growth of civilization … The objective of restorative environmental design depends on wooing the earth in a deliberate, knowing, and gentle fashion".*
>
> (Kellert, 2005, 22)

From an EE perspective, care can work as an emotional impetus for nature protection but is ambiguous as a rationale for actions towards nature. Johns-Putra explains, for example, that care for nature is not *"the means by which agency occurs; it is itself agential"* (2013, 126). In other words, "care" cannot be regarded as a motivator for an action towards something, but rather as a form of relational agency between a caregiver and a caretaker. Attentive caring for nonhuman life implies then that nature is "responding" to human actions as if it were a "caretaker". Likewise, the relativity and contingency of care for nature become evident when considering encounters with specific nature phenomena – very few would, for example, exercise care for a virus.

The notion of "care for nature" functions in biophilic concepts as a "meta narrative" (Lyotard, 1984, xxiii)[16] rather than providing a reasonable principle for moral maxims and it may be argued that ethical implications of care become more convincing when grounded in lived experiences of nature and phenomenology as for example in Heidegger (1996), see also Chapter 3.4.1.

In design, Dilnot's (2011) analysis of the underlying causes of unsustainability bears notable resemblance to principles found in deep ecology. He argues that the relationship between humans and nature must be fundamentally reimagined and reconciled. In a provocative and controversial analogy, Dilnot asserts: *"the unsustainable is no more a problem of nature than was the 'Jewish problem' a problem of the Jews"* (ibid., p. 103). Through this comparison, Dilnot characterises unsustainability as a human-made crisis and places collective responsibility on humanity for its failure to live in harmony with the natural world.

Living in a *"catastrophic lack of sustainable relations between economic activity, technics and nature"* (ibid., 119) is society's fault. Nature, on the other hand, is perceived as wishing humans well, while unhappiness is humans' own fault, because they are violating the principle of *"hospitality towards all"* (ibid., 104). These assumptions align with key principles of deep ecology, particularly in their emphasis on a binary opposition between a benevolent nature and destructive human activity, scepticism towards technological solutions, and the call for reconciliation between humanity and the natural world.

Dilnot outlines nine conditions for sustainment and articulates his normative stance in his seminal work *Ethics in Design: 10 Questions* (2009). He begins by asserting a fundamental ethical responsibility that underpins sustainable practice. This includes acknowledging human flaws that result in impeding conditions of lives, *"ours as much as others"* (ibid., 135). A remedy is a strong ethic of sustainability that eliminates false and unnecessary needs and attempts to reverse the unsustainable. While Dilnot's ethics remains vague regarding socio-political discourses, he suggests a germane argument for sustainable design ethics by employing aesthetic means to acknowledge ethical challenges and possibilities in human–nature relationships, regarding the responsibility of the designer and the role of products:

> *"Design is the discovery of what the artificial can be for us. Since the artificial is also today the frame of our possibilities as human beings, to discover what the artificial can be for us is to discover what our possibility can be and hence … it is also a discovery of what possibility can be".*
>
> (Dilnot, 2009, 84)

Design is presented as potentially valuable for societies. Dilnot sees its aesthetic potential as especially beneficial for connecting technical-instrumental, interactive-communicative, and subjective-experiential components, providing a *"concrete model of how these apparently antithetical realms of human experience can be brought together"* (Dilnot, 1982, 22).

Design as a discipline and practice is relevant either in materialising or excluding ethics (Fry et al., 2003). Considering the discussion above, deep ecology as environmental philosophy appears a problematic companion for sustainable design and planning ethics. However, ideas such as "aesthetic *telos*" and aesthetic connectivity as reconciliation of the instrumental, the communicative and the phenomenological are beneficial for an enlarged SD debate. Most importantly, deep ecology rejuvenates the insight that empirical, normative, and lifeworldly nature interpretations are all part of sustainability discourses. While the possibility to generalise nature experiences allows efficiency in the empirical sciences, it also separates humans from the subjective and meaningful elements of these experiences. The natural sciences play an indispensable role in the evaluation of technical processes and in the discussion of environmental issues, which not least relates to their ability to warn of hazards and to provide remedial action in ecologically critical situations. Yet, facts and models provide relatively limited access to reality and cannot claim to represent the totality of life experiences. According to deep ecology, it would therefore be ruinous to completely adopt the scientific worldview, because it contains the renunciation of sensory-subjective experiences, which are just as essential for communication about nature as empirical descriptions.

The following chapter addresses the challenge of analysing and systematising a set of concepts that differ significantly in how they engage with sustainable design. These concepts present contrasting views on human–nature relationships, and the responsibilities of professionals. They also diverge in how they define the purpose of products and services, as well as in their underlying value systems, ethical imperatives, and normative expectations for designers.

Notes

1 In brief, critical anthropomorphism attributes emotions and intentions to non-human beings, while recognising the anthropocentric perspective of these projections. Avoiding simplistic or uncritical attributions, it acknowledges that nature attributions contribute to episteme and ethos, and aims to foster nuanced interpretations (Guthrie, 1993). A more detailed discussion of this notion can be found in Chapter 5.1.

2 The antinomy between human autonomy and determination by nature is a long-standing debate in philosophy. Hegel, Marx, and Foucault, for example, argue that human actions are never fully autonomous but are always shaped by natural, social, or historical conditions. While this perspective is convincing, the discussion of autonomy in this book draws on Kant's concept of negative freedom, which defines autonomy not as total independence from all influences but as freedom from external coercion (*Groundwork of the Metaphysics of Morals*, 1993). Even if absolute autonomy is unattainable, according to Kant, individuals may still experience themselves as autonomous through moral reflection, ethical action, and aesthetic engagement. Nature's autonomy is further reflected in Adorno's concept of the non-identical and Heidegger's earth–world dialectic, both of which resist the reduction of nature to a purely conceptual or instrumental object. The "non-identical" and the "earth" each preserve inaccessible, experiential, and material dimensions of nature, maintaining its autonomy as something more than a

resource or object of control. These ideas will be explored in more depth in the chapters that follow.

3 Früchtl (1996), original quote: "Ein und derselbe Gegenstand läßt ganz verschiedene Stimmungen zu. Die Menschen haben verschiedene Biografien, und wo der eine sich angesichts eines Gartens, in dem knallrote Tulpen blühen und die Rotkehlchen zwitschern, an glückliche Kindertage erinnert fühlt, muß der andere daran denken, wie in einem solchen Garten einst vor seinen Augen der Tante die Kehle durchgeschnitten wurde" (ibid., 403).

4 Moholy-Nagy (1947) had already in 1947 advocated for the integration of ethics in design, though not explicitly in relation to environmental concerns.

5 Philosophically the "empirical turn" from the 1950s relates to a critique of technology pessimism and the claim that in order interpret technologies, one must study them empirically. Here the term is used to characterise the shift from artistic and handicraft-oriented design to engineering-oriented industrial design.

6 Narrative turn connotes here a move in design to storytelling (Barthes et al., 1975) and hermeneutic interpretation of artefacts, supplementing technical and functional design approaches. Narratives are discussed more detailed in Chapter 4.

7 Tonkinwise, C. (2003): *"here is a clear account of our societies' unsustainability. Our 'taking for granted what we have made because we have made it seem complete', allows us to proliferate our world with things, some disappearing in use, but most just hanging around disused, persisting in unproductive or polluting ways. We are then literally unsustainable, unable to sustain all that we scatter about ourselves"* (ibid., 9). http://www.ub.edu/5ead/PDF/8/Tonkinwise.pdf. Retrieved 31 January 2023.

8 In this book, "emancipatory" refers to individuals' rights to participate in and actively shape political and social processes.

9 Merriam Webster defines psychotechnology as: "the application of psychological methods and results to the solution of practical problems especially in industry". https://www.merriam-webster.com/medical/psychotechnology. Retrieved 31 January 2023.

10 The distinctions represent extreme positions and are intended solely for analytical purposes. De facto, many SSD concepts move in a grey zone between psychotechnical and emancipatory approaches.

11 Conceptually, both social constructivism and social constructionism are grounded in the idea that knowledge is not objective but rather co-constructed through human interaction. Reality is understood as socially produced by actors, and meaning emerges through individuals' engagement within these interactional frameworks. The notion that phenomena possess no inherent meaning independent of human mental or linguistic interpretation is further developed in postmodern thought, which challenges the existence of fixed or universal truths.

12 Meliorism believes that progress can be produced by humans, and that it can enhance natural processes. Ethically, meliorism relates to utilitarism and pragmatism (Payton and Moody 2008). Pragmatism evaluates the validity of theories (and ethics) regarding the success of their practical application.

13 Analytical engineering science approaches to locate possible responses to moral problems provide sufficient and comprehensive introductions for industrial designers but are often deficient in terms of analysing and implementing sustainable design values, e.g., in Harris et al. (2009): *"The virtue of respect for nature is important in motivating engineers to engage in environment-friendly engineering. It can be cultivated by a more intellectual and scientifically based appreciation of the value to humans of the experience of the natural world and by the direct experience of the transcendent aspect of nature"* (ibid., 173).

14 While *"antinomy signifies a contradiction between positions (humans have a free will and humans are determined by nature), dichotomy means the exclusion of positions because they are entirely different ('sustainability' and 'overconsumption')".*

15 Aesthetic telos can be connoted "experienced purposefulness" (Keitsch, 2002b). For an in-depth discussion of this notion, see Chapters 3.7 and 5.1 of this issue.

16 The terms *metanarrative* and *grand narrative* are often used interchangeably, especially in postmodernism, to describe overarching ideas that explain and give meaning to other ideas. Metanarratives typically assert a stronger claim to universal truth and explanatory authority. In biocentrism, for example, the idea of nature's intrinsic value functions as a metanarrative embedding the human–nature relationship within an universal moral structure. The Enlightenment metanarrative claims that knowledge and science will lead to freedom and progress. The *ex negativo* postmodernist metanarrative asserts that all narratives are limited while paradoxically making that assertion universally.

References

Adorno, T. W. (2001). *Negative dialectics* (D. Redmond, Trans.; rev. ed. 2021). https://platypus1917.org/wp-content/uploads/Negative_Dialectics_Redmondtrans2021.pdf. Retrieved 28 December 2022.

Adorno, T. W. (1997). *Aesthetic theory* (R. Hullot-Kentor, Trans.). Continuum.

Aiken, W. (1984). Ethical issues in agriculture. In T. Regan (Ed.), *Earthbound: New introductory essays in environmental ethics.* Random House.

Airenti, G. (2018). The development of anthropomorphism in interaction: Intersubjectivity, imagination, and theory of mind. *Frontiers in Psychology*, 9, Article 1835.

Andersen, H., & Sørensen, K. (1994). *Frankenstein's dilemma*. Ad Notam Gyldendal.

Anker, P. (1999). From scepticism to dogmatism and back. In N. Witoszek & A. Brennan (Eds.), *Philosophical dialogues: Arne Næss and the progress of eco-philosophy.* Rowman & Littlefield.

Barthes, R., & Duisit, L. (1975). An introduction to the structural analysis of narrative. *New Literary History*, 6(2), 237–272.

Bassey, S. (2020). Technology, environmental sustainability and the ethics of anthropoholism. In *Proceedings of the International Symposium for Environmental Science and Engineering Research (ISESER 2020)*. https://socialspacejournal.eu/20%20numer/Technology,%20environmental%20sustainability%20-%20Bassey.pdf. Retrieved 28 December 2022.

Beatley, T. (2011). *Biophilic cities: Integrating nature into urban design and planning.* Island Press.

Bentham, J. (1789). *An introduction to the principles of morals and legislation* (Chap. 17). https://www.earlymoderntexts.com/assets/pdfs/bentham1780.pdf. Retrieved 20 December 2025.

Benyus, J. M. (2002). *Biomimicry: Innovation inspired by nature.* Harper Perennial.

Bethelmy, L., & Corraliza, J. A. (2019). Transcendence and sublime experience in nature: Awe and inspiring energy. *Frontiers in Psychology*, 10, Article 509.

Bhamra, T., Lilley, D., & Tang, T. (2011). Design for sustainable behaviour: Using products to change consumer behaviour. *The Design Journal*, 14(4), 427–445.

Bigell, W. (2007). Cultural significations of heterotopian spaces of nature. *American Studies in Scandinavia*, 39(2), 1–13.

Bonsiepe, G. (2006). Design and democracy. *Design Issues*, 22(2), 27–34.

Bookchin, M. (1999). *Social ecology versus deep ecology: A challenge for the ecology movement*. Rowman & Littlefield.

Brandi, C. (2015). Safeguarding the earth system as a priority for sustainable development and global ethics: The need for an earth system SDG. *Journal of Global Ethics*, 11(1), 32–36.
Brey, P. (2006). Ethical aspects of behaviour-steering technology. In P.-P. Verbeek & A. Slob (Eds.), *User behavior and technology development*. Springer.
Brunello, L. (2025). What if service design meets deep ecology? Implications on ecosystemic services opportunities in urban systems. https://www.researchgate.net/publication/394309578_What_if_Service_Design_meets_Deep_Ecology_Implications_on_Ecosystemic_Services_Opportunities_in_Urban_Systems. Retrieved 28 December 2022.
Caperna, A. (Ed.). (2018). Biophilic design. *Journal of Biourbanism*, 6(1–2). International Society of Biourbanism.
De Medeiros, J. F., da Rocha, C. G., & Ribeiro, J. L. D. (2018). Design for sustainable behavior (DfSB): Analysis of existing frameworks of behavior change strategies, experts' assessment and proposal for a decision support diagram. *Journal of Cleaner Production*, 188, 402–415.
Dilnot, C. (1982). Design as a socially significant activity: An introduction. *Design Studies*, 3(1), 39–146.
Dilnot, C. (2009). Ethics in design: 10 questions. In H. Clark & D. Brody (Eds.), *Design studies: A reader*. Berg.
Dilnot, C. (2011). Sustainability and unsustainability in a world become artificial: Sustainability as a project of history. *Design Philosophy Papers*, 9(2), 103–155.
Drenthen, M. (2017). Environmental hermeneutics and the meaning of nature. In A. Thompson & S. Gardiner (Eds.), *The Oxford handbook of environmental ethics*. Oxford University Press.
Dreyfus, H. L. (1967). *The measure of man: Human factors in design*. Whitney.
Ehrenfeld, J. R. (2000). Industrial ecology: Paradigm shift or normal science? *American Behavioral Scientist*, 44(2), 229–244.
Ehrenfeld, J. R. (2008). *Sustainability by design*. Yale University Press.
Enzensberger, H. M. (1974). A critique of political ecology. *New Left Review*, 84, 3–31.
European Union. (2008). Sustainable consumption and production and sustainable industrial policy action plan. https://www.eea.europa.eu/policy-documents/com-2008-397 Retrieved 28 December 2024.
Fathers, J. (2003). Peripheral vision: An interview with Gui Bonsiepe. *Design Issues*, 19 (4), 44–56.
Findeli, A. (2001). Rethinking design education for the 21st century: Theoretical, methodological, and ethical discussion. *Design Issues*, 17(1), 5–17.
Fogg, B. J., & Hreha, J. (2010). Behavior wizard: A method for matching target behaviors with solutions. In *Persuasive technology (Lecture Notes in Computer Science*, Vol. 6137, pp. 117–131). Springer.
Fraser-Darling, F. (1969). Man's responsibility for the environment. In F. J. Ebling (Ed.), *Biology and ethics*. Academic Press.
Franck, G. (2018). The economy of attention. *Journal of Sociology*, 55(1), 6–18.
Frey, R. G. (1989). Why animals lack beliefs and desires. In P. Singer & T. Regan (Eds.), *Animal rights and human obligations*. Prentice Hall.
Frosch, R. A., & Gallopoulos, N. E. (1989). Strategies for manufacturing. *Scientific American*, 261(3), 94–102.
Früchtl, J. (1996). *Ästhetische Erfahrung und moralisches Urteil*. Suhrkamp.
Fry, T. (2009). *Design futuring: Sustainability, ethics and new practice*. Berg.

Fry, T., & Dilnot, C. (2003). Manifesto for redirective design: Hot debate. *Design Philosophy Papers*, 1(2), 95–103.

Fuad-Luke, A. (2009). *Design activism*. Earthscan.

Gadamer, H.-G. (1986). *The relevance of the beautiful and other essays* (N. Walker, Trans.; R. Bernasconi, Ed.). Cambridge University Press.

Giri, A. K. (2002). The calling of a creative transdisciplinarity. *Futures*, 34(1), 103–115.

Guthrie, S. (1993). *Faces in the clouds: A new theory of religion*. Oxford University Press.

Habermas, J. (1974). *Theory and practice* (J. Viertel, Trans.). Heinemann.

Harris, C. E., Pritchard, M. S., & Rabins, M. J. (2009). *Engineering ethics: Concepts and cases* (4th ed.). Cengage.

Heidegger, M. (1996). *Being and time* (J. Stambaugh, Trans.). SUNY Press.

Heskett, J. (1980). *Industrial design*. Thames & Hudson.

Hung, R. (2017). Towards ecopedagogy: An education embracing ecophilia. *Educational Studies in Japan: International Yearbook*, 11, 43–56.

Isenmann, R., Bey, C., & Keitsch, M. (2007). Beyond a sack of resources: Nature as a model – core feature of industrial ecology. In M. Ruth & B. Davidsdottir (Eds.), *Changing stocks, flows and behaviors in industrial ecosystems*. Edward Elgar.

Johns-Putra, A. (2013). Environmental care ethics: Notes toward a new materialist critique. *symplokē*, 21(1–2), 125–135.

Jonas, H. (1996). *The imperative of responsibility: In search of ethics for the technological age*. University of Chicago Press.

Kant, I. (1993). *Groundwork of the metaphysic of morals* (J. W. Ellington, Trans.; 3rd ed.). Hackett Publishing.

Keitsch, M. (2002a). *Naturästhetik und ökologische Ethik: Eine Einführung*. Verlag Dr. Kovač.

Keitsch, M. (2002b). Some perspectives on environmental ethics. *Parabel: Tidsskrift for filosofi og vitenskapsteori*, 30–35.

Keitsch, M. (2012). Ethical implications of systems methodology applied to industrial ecology. *Sociology Study*, 2(1), 49–59.

Keitsch, M. (2015). Sustainable design: Concepts, methods, and practices. In M. Redclift & D. Springett (Eds.), *Routledge international handbook of sustainable development*. Routledge.

Keitsch, M., & Bjørnstad, N. (2010). Ethics in product design curriculum: An example from the Oslo School of Architecture and Design. In *Proceedings of the 12th International Conference on Engineering and Product Design* (NTNU, Trondheim).

Kellert, S. R., & Wilson, E. O. (1995). *The biophilia hypothesis*. Island Press.

Kellert, S. R. (2005). *Building for life: Designing and understanding the human–nature connection*. University of Chicago Press.

Krebs, A. (1999). *Ethics of nature: A map*. De Gruyter.

Krippendorff, K. (1989). On the essential context of artifacts or on the proposition that "design is making sense (of things)". *Design Issues*, 5(2), 9–39.

Lloyd, P. (2002). Making a drama out of a process: How television represents designing. *Design Studies*, 23(2), 113–133.

Lockton, D., Harrison, D., & Stanton, N. A. (2008). Making the user more efficient: Design for sustainable behaviour. *International Journal of Sustainable Engineering*, 1 (1), 3–8.

Lyotard, J.-F. (1984). *The postmodern condition: A report on knowledge*. University of Minnesota Press.

MacIntyre, A. (1981). *After virtue: A study in moral theory.* University of Notre Dame Press.
Macy, J. (2007). *World as lover, world as self: Courage for global justice and ecological renewal.* Parallax Press.
Madge, P. (1997). Ecological design: A new critique. *Design Issues*, 13(2), 44–54.
Manzini, E. (2003). Scenarios of sustainable well-being. *Design Philosophy Papers*, Issue 1, 5–21.
Manzini, E. (2006). Design, ethics and sustainability: Guidelines for a transition phase. *Cumulus Working Papers.* University of Art and Design Helsinki.
Manzini, E., & Cullars, J. (1992). Prometheus of the everyday: The ecology of the artificial and the designer's responsibility. *Design Issues*, 9(1), 5–20.
Manzini, E., & Jégou, F. (2003). *Sustainable everyday: Scenarios of urban life.* Edizioni Ambiente.
Marshall, D. (2021). The commodified celebrity-self: Industrialized agency and the contemporary attention economy. *Popular Communication*, 19(3), 164–177.
McGregor, S. L. T. (2015). Transdisciplinary knowledge creation. In P. T. Gibbs (Ed.), *Transdisciplinary professional learning and practice.* Springer.
McLennan, J. F. (2004). *The philosophy of sustainable design.* Ecotone.
Miller, P. (1982). Value as richness: Toward a value theory for an expanded naturalism in environmental ethics. *Environmental Ethics*, 4, 101–114.
Moholy-Nagy, L. (1947). *Vision in motion.* Paul Theobald.
Nelson, E. (2011). Revisiting the dialectic of environment: Nature as ideology and ethics in Adorno and the Frankfurt School. *Telos*, 155, 105–126.
Næss, A. (1989). *Ecology, community, and lifestyle: Outline of an ecosophy.* Cambridge University Press.
Naess, A., & Drengson, A. R. (2008). *Ecology of wisdom: Writings by Arne Naess.* Counterpoint Press.
Odum, E. P. (1969). The strategy of ecosystem development. *Science*, 164(3877), 262–270.
Omotosho, I. F. (2019). Between utility and rights: The expected changes in Tom Regan's defence of the right of animals. *Journal of Philosophy, Culture and Religion*, 40, 4–8.
Papanek, V. (1985). *Design for the real world: Human ecology and social change.* Academy Chicago Publishers.
Papanek, V. (2021). *The green imperative.* Thames & Hudson.
Patterson, M. (2006). Consumption and everyday life. In A. Elliott (Ed.), *The new sociology.* Routledge.
Payton, R. L., & Moody, M. P. (2008). *Understanding philanthropy: Its meaning and mission.* Indiana University Press.
Pettersen, I. N., & Boks, C. (2008). The ethics in balancing control and freedom when engineering solutions for sustainable behaviour. *International Journal of Sustainable Engineering*, 1(4), 287–297.
Pichler, M., Schaffartzik, A., Haberl, H., & Görg, C. (2017). Drivers of society–nature relations in the Anthropocene and their implications for sustainability transformations. *Current Opinion in Environmental Sustainability*, (26–27), 32–36.
Rawls, J. (1971). *A theory of justice.* Harvard University Press.
Regan, T. (1986). A case for animal rights. In M. Fox & L. Mickley (Eds.), *Advances in animal welfare science 1986/87.* The Humane Society of the United States.
Rolston, H. (1991). Environmental ethics: Values in and duties to the natural world. In H. Bormann & S. R. Kellert (Eds.), *The broken circle: Ecology, economics, ethics.* Yale University Press.

Salingaros, N. A. (2000). Complexity and urban coherence. *Journal of Urban Design*, 5, 291–316.
Sessions, G. (1995). *Deep ecology for the 21st century: Readings on the philosophy and practice of the new environmentalism*. Shambhala.
Shiva, V. (2016). *The violence of the Green Revolution: Third World agriculture, ecology, and politics*. University Press of Kentucky.
Shove, E. (2003). *Comfort, cleanliness and convenience: The social organisation of normality*. Berg.
Shu-Yang, F., Freedman, B., & Côté, R. (2004). Principles and practice of ecological design. *Environmental Review*, 12, 97–112.
Singer, P. (1990). *Animal liberation* (Rev. ed.). Review Books.
Spaemann, R. (1980). Technische Eingriffe in die Natur als Problem der politischen Ethik. In D. Birnbacher (Ed.), *Ökologie und Ethik*. Reclam.
Spaemann, R. (2007). Art imitates nature? In D. C. Schindler & J. H. Schindler (Eds.), *A Robert Spaemann reader* (pp. 192–210). Oxford University Press.
Spaemann, R. (2012). Human dignity and human nature. In R. Spaemann & D. Schindler (Eds.), *Love and the dignity of human life: On nature and natural law*. Eerdmans.
Stanford Encyclopedia of Philosophy (20025), Epistemology, https://plato.stanford.edu/entries/epistemology/. Retrieved 6 July 2026.
Taylor, P. W. (1981). The ethics of respect for nature. *Environmental Ethics*, 3(3), 197–218.
Thackara, J. (2005). *In the bubble: Designing in a complex world*. MIT Press.
Tonkinwise, C. (2003). Interminable design: Techne and time in the design of sustainable service systems. http://www.ub.edu/5ead/PDF/8/Tonkinwise.pdf. Retrieved 31 January 2023.
Tran, B. (2017). Psychological (and emotional) architecture: The values and benefits of nature-based architecture biophilia. In *Cultural influences on architecture*. IGI Global.
Unger, D. W., & Eppinger, S. D. (2011). Improving product development process design: A method for managing information flows, risks, and iterations. *Journal of Engineering Design*, 22(10), 689–699.
United Nations. (1995). World summit for social development (Annex 1). http://www.un.org/documents/ga/conf166/aconf166-9.htm. Retrieved 30 January 2023.
United Nations Environment Programme. (2012). Cleaner production promotion law. https://leap.unep.org/countries/cn/national-legislation/cleaner-production-promotion-law-2012. Retrieved 20 December 2025.
United Nations Environment Programme. (2015). Sustainable Development Goal 12. https://www.unep.org/explore-topics/sustainable-development-goals/why-do-sustainable-development-goals-matter/goal-12. Retrieved 20 December 2025.
United Nations General Assembly. (2015). Resolution adopted by the General Assembly on 25 September 2015 (A/RES/70/1).
Van Berkel, R. (2000). Cleaner production perspectives for the next decade. In *Proceedings of the 6th international high-level seminar on cleaner production* (Montreal, Canada, 16–17 October).
Van der Ryn, S., & Cowan, S. (2007). *Ecological design* (10th anniversary ed.). Island Press.
Van Doren, H. (1954). *Industrial design: A practical guide to product design and development* (2nd ed.). McGraw-Hill.
Vetlesen, A. J. (2015). *The denial of nature: Environmental philosophy in the era of global capitalism*. Routledge.
Zimmerman, M. E. (1995). The threat of ecofascism. *Social Theory and Practice*, 21 (2), 207–238.

3 Recent Sustainable Design Concepts

3.1 Design for Sustainability Transitions

Progressing towards sustainable development entails balancing economic growth, social inclusion, and environmental protection, and the Sustainable Development Goals (SDGs) offer a universal framework to support this progress. Some authors claim, however, that SDG advancement remains limited, especially on a global scale (Sachs et al., 2019; Leal Filho et al., 2020) and that complex societal and environmental challenges such as climate change, biodiversity degradation, and air and water pollution are particularly difficult to address due to their systemic entanglement. Other authors emphasise that capitalist ideologies and unsustainable patterns of production and consumption are key drivers of both social and environmental stress, posing significant barriers to the achievement of the Sustainable Development Goals (Kemp et al., 2018; Feola, 2020). In the consumption context, the role of individuals is increasingly recognised as critical. Individuals are not only interpreted as consumers but also citizens, workers, voters, and community members who can bring about change, particularly when supported by enabling environments and political stakeholders. Moreover, the attitudes and values of citizens are seen as pivotal in either facilitating or hindering sustainable transformations and lifestyle changes (Akenji et al., 2016).

Within this setting, sustainability transitions (ST) aspire to initiate systemic change on either socio-technical, socio-ecological, or socio-economic levels[1]. Sustainability transitions are characterised as complex, systemic, and multi-faceted processes of structural changes in society. Grin et al. (2010) describe ST as *"radical transformation towards a sustainable society, as a response to a number of persistent problems confronting contemporary modern societies"* (ibid., 1). A central focus within the field is the development of strategies and instruments to direct socio-technical systems towards more sustainable futures (Geels, 2005), while concepts do not only emphasise technological innovation but also transformations in socio-cultural norms, behavioural patterns, institutional structures, and organisational practices (Köhler et al., 2019).

Sustainability transition research has existed since the 1980s and has evolved into a transdisciplinary field comprising various interpretations,

DOI: 10.4324/9781003384090-3

disciplinary views, and methodologies. ST concepts encompass, among others, resilience theory, decoupling theory, the green economy and green growth, behavioural change theory, and the Transition Towns movement (Luederitz et al., 2017).

Ethical considerations in sustainability transitions (ST) manifest either implicitly or explicitly through deliberate normative arguments. Implicit ethics in ST concepts incorporate classical moral values such as justice, equity, and the fair distribution of resources (Köhler et al., 2019). Ethical dimensions are typically assumed rather than problematised, operating beneath the surface of technological and institutional change. By contrast, normative ethics seek to explicitly articulate and develop guiding principles for norms, maxims and actions. These approaches are more deliberate in their ethical engagement, as seen in emerging perspectives such as eco-spirituality, which aim to reframe sustainability through value-laden worldviews and moral imperatives (Lestar et al., 2020).

"Justice" is a core socio-technical ST value and norm, often associated with individual "contractualism"[2] (Huntjens, 2021). A contract model is an agreement of autonomous individuals who commonly make decisions, which necessitates that decisions and actions such as sustainability innovations in society can be initiated by governments or other societal stakeholders but must be approved and supported by the majority of the population or stakeholders (Avelino, 2017). Contractualism holds the anthropocentric premise that human interactions imply mutual moral responsibility, and ST contractualist approaches benefit social sustainability through pursuing democratic values. ST authors further analyse power relations, styles of power, and power dynamics, here described as interactions of different stakeholders with resources, structures, and systems, and considering the domination these stakeholders exert on each other within contractual limitations (Avelino, 2017):

> *"Institutions enforcing environmental regulations would only be effective if they are morally supported by a majority of the community, and if not, such an enforced order could even have counterproductive effects in terms of decreasing the effectiveness of power exercise".*
>
> (ibid., 514)

Criticism of contractual ST addresses, among others, the assumption that societal and individual values and principles correlate and are sometimes treated as equal in ST concepts. Spahn (2016), for example, presents an epistemic critique that challenges the assumption of an idealised view of all societal members as free and rational agents striving for consensus and justice in socio-political matters, whereas in reality, decision-making is often driven by power dynamics and irrational factors[3].

A semantic critique of socio-technical sustainability transition (ST) concepts argues that these often conceal underlying ideologies. While analyses of structures, conditions, and strategic interventions may appear neutral, they frequently rest on unacknowledged normative assumptions such as adherence to a liberal market paradigm (Van Assche et al., 2022). Governments are typically portrayed as mediating agents that soften the effects of capitalist systems, yet they may also contribute to the problem. As early as 1974, Enzensberger warned of demagogic intentions behind the ecological movement and noted that the concept of "sustainable development" was initially shaped by top managers and bureaucrats:

> *"the monopolies attempt to acquire influence over the ecological movement. The MIT study commissioned by the Club of Rome is by no means the only initiative of this kind. The monopolies are also represented in all state and private commissions on the protection of the environment. Their influence on legislation is decisive, and there are numerous indications that even apparently spontaneous ecological campaigns have been promoted by large firms and government departments".*
>
> (Keitsch, 2021, 12)

Considering tacit ideological and normative underpinnings of sustainability transition concepts, recent authors such as Shove et al. (2007) argue that ST methodologies cannot be considered "neutral". *"Sustainability is tacitly defined as a matter of resource management, efficiency, and ecological modernisation"* (Shove et al. 2007, 768). *"Even the most primitive attempt to establish starting assumptions (on sustainable societies, MK) would immediately reveal divisions and fractures between opposing interests and ideologies"* (Shove et al. 2007, 766). Shove et al. (2007) further address the structural challenge to review former, current, and evolving cultural and political values of sustainability and their effects on ST management and design. Besides querying the generic interpretation of citizens, criticisms of ST norms frequently relate to employing capitalist politics and technocracy as standardising fundamentals and to excluding race- and gender-related issues. Some authors suggest "just transitions" as an alternative to ST. Rather than merely discussing ST through management or policy planning, these authors focus on eliminating negative side effects of industrial capitalism by improving practices in societies and developing new inclusive ways of social and design innovation, for example through design futuring (Newell et al., 2013; Whyte, 2018; White, 2020). However, just transitions concepts appear more fragmented than ST theories since they can relate to various frameworks such as labour, justice, socio-technical transition, governance, and public perception (Wang et al., 2021).

Ethically, criticism is further directed towards the anthropocentric postulate of a "healthy environment" for humans (Knox et al., 2018) and ST's overly instrumental focus on nature (Spahn, 2016). In contrast to technocratic and bureaucratic ST concepts, eco-spirituality offers a normative ethical interpretation of sustainability transitions in diverse cultural contexts and for distinct individual lifestyles, proposing the conditionality of moral choices. According to Koehrsen et al. (2021), spirituality and religious beliefs comprise visions and moral attitudes that can regulate overly instrumental and operational features of ST. Lestar et al. (2020) point to the lacking conception of agency in ST concepts and emphasise the social value of religion and spirituality, asserting that pro-environmental values and practices are generally more successfully maintained in spiritual than in secular settings. Eco-spiritual communities and religious movements often embrace values such as simplicity, frugality, and modesty as pathways toward sustainability, as illustrated in subchapter 4.3.2 of this book, and may play a significant role in facilitating sustainability transitions in everyday contexts. Lestar et al. (2020) argue further that studying social practices and lived experience of practitioners, and paying attention to social phenomena such as spirituality will not only uncover which beliefs as fundament of current practices are dominating and why societal actors support them, but also open possibilities for radical changes in sustainable transition research and development.

Epistemically, this practice interpretation corresponds with a change from subject–object oriented to intersubjective world understanding: *"Everybody is the other and no one is himself"* (Heidegger, 1962, 165), which connects human existence to countless possibilities for experiencing and interacting with the surroundings. Empirical practice elements such as routinised behaviour, bodily and mental activities, artefacts, and their use, etc., are here surpassed by the assumption of a teleoaffective relational structure of humans and the material world, as e.g., Friedland (2018, 1359) points out, *"Modalities of practice … offer a way to restore value not only as a reason located in the subject, as embodied in the object, but as carried by practice"*. Teleoaffective reasoning emphasises emotional and experience values orientation of human–nature interactions, beyond instrumental value. While the internal logic of this notion may be subject to philosophical debate, it holds normative significance by framing environmental engagement not merely as utility-driven, but as grounded in meaning and intentionality. Teleoaffectiveness is relevant to both environmental ethics and sustainable design, where the motivation to act often arises not only from rational decision-making or trade-offs, but also from values and emotional commitments toward the natural and social world in concert.[4]

Design for sustainable transitions and transition design[5] aim both to harmonise lifestyles with natural environments and sustainable futures, exploring possibilities to meet human needs within networks of local resilience (Irwin, 2015). Transition design suggests a threefold ethical procedure for professionals: to reflect on one's value mindset, to take responsibility for decisions, and to implement solutions holistically and collaboratively (Irwin et al., 2015).

Methodologically, transition designers apply scenarios, design futuring, narratives, and storytelling and claim that visions of sustainable futures can inform present design solutions. Economically, equitable, alternative modes of exchange are seen as an important cornerstone for sustainable transitions (Mander, 2012). Transition design concepts call for interpreting and articulating sustainability ideas and value propositions, e.g. by *"asking designers to examine their own value system and the role it plays in the design process. … transition solutions will be best conceived within a more holistic worldview that can inform new, more collaborative, and responsible postures for interaction"* (Irwin, 2015, 235). Transition towns, where local food and energy systems, alternative currencies, and novel products and services are developed, are a prominent practical design and planning example (Hopkins, 2008).

Yet, transition design concepts hardly discuss which values should be articulated, negotiated, embraced, and eventually realised, and sustainability interpretations in the concepts often remain vague. Some ST authors suggest environmental ethics as a value framework but simultaneously propose feminist theory, animal studies, degrowth studies, post-humanist ethnography, political ecology, and literature on decolonising methodology as sources for value generation (Gaziulusoy et al., 2019). Others define value questions as situational, based on practice theory (Friedland, 2018). Without establishing a value canon, Mok et al. (2018) provide a useful methodological approach by aligning a value-sensitive design framework (VSD) with situational ethics[6]. Universal ethical values related to environmental sustainability, such as the common good and justice, are here connected to situative, empirical investigations. Mok et al. (2018) further manage to show value conflicts and trade-offs due to context dependency and differences in value perceptions. When ST is reflected as a design challenge in concepts, a major task of a transition designer is to enhance everyday life of users and stakeholders by restoring relationships between people, nature, and artefacts (Gaziulusoy et al., 2018). Alternatively, a relativist position sees the binary view of professional designers and laypersons as obsolete (Heiskanen et al., 2014) and claims that design agency for all, instead of fixed roles, should be emphasised in bottom-up transitions. Co-creation for ST is one way to address grassroots and local levels. It is part of participatory design approaches, which are discussed in Chapter 4.

3.2 Participatory Design and Social Sustainability

The environment as a core arena for stakeholder collaboration and user inclusion[7] is mentioned early in sustainability discourses, e.g., in the *Report of the World Commission on Environment and Development – Our Common Future* (World Commission of Environment and Development, 1987), which stated that political systems should ensure participation in public decisions to achieve sustainable development. This statement has evolved into a mantra over the last 35 years, and national and international organisations commonly refer to public or stakeholder participation as a way of including various actors in planning and implementing projects and political agendas (Lohse et al., 2023). Hickey and Mohan (2004) estimate that participatory approaches as tools for good governance entered mainstream political and social practices in the mid-1980s. Today, many governments and government agencies have embraced participatory approaches, for example within "citizen engagement"; participation has become an asset for policymaking (Huttunen et al., 2022), and authors commonly agree that participation leads to greater inclusion of marginalised groups (Herrle et al., 2012).

Undertaking participation in design or planning processes serves many purposes, from discovering issues, searching for problem alternatives, educating the public and facilitating learning, discussing and deciding on a solution, measuring public opinion, legitimising public decisions, including stakeholders in development processes, maintaining social connections, and bridging conflict situations. Correspondingly, participatory design (PD) is increasingly recognised in public and organisational settings (Pilemalm, 2018). The aim of PD is to identify and manage different stakeholders' interpretations, needs, and skills (Sanders et al., 2008) for both implementation of policies on local levels, and planning, development, and implementation of solutions to specific problems. PD explicitly emphasises social inclusion, regarding stakeholders as partners in the entire design process, from problem definition to idea generation, information collection, solution suggestion, decision-making, specification, implementation, and finally the evaluation phase. Some authors claim that participatory design fosters political competence, increases interest in political processes, builds trust between societal stakeholders, and strengthens the belief in transforming practice. Participatory design perceives stakeholders as experts of their experience (Visser et al., 2005) and aims to connect different types of knowledge (Araki, 2021), thereby enabling co-production and generating partnerships among social stakeholders. Opposed to classical participation approaches and user-centred problem-solving processes, where experts are the organisers, conductors, and creators, participatory design sees experts or professionals as facilitators providing tools that allow collaboration in a team (Sanders et al., 2008).

The level of stakeholder inclusion in participatory design approaches varies theoretically as well as in practice. At a minimum level, stakeholders are included through, e.g., reporting problems to experts and professionals, who will develop solutions which are then tested by the stakeholders. At a medium level, inclusion comprises stakeholder engagement in the form of, for example, constructive suggestions in planning, development, and implementation phases, but stakeholders are not part of the decision-making process (Mogstad et al., 2018). At a maximum level, stakeholders become co-designers (Sanders et al., 2008), and project facilitators will provide time, permission, space, and tools for co-creation activities. At this level, the chance for stakeholders to experience that they are working on something personally important to them or their group, and that their contribution has an impact, is relatively high.

In practice, participatory design faces, among others, the issue of whether stakeholders feel like they are part of a team or are just being consulted randomly. According to Bowen et al. (2013), many project participants experience de facto being an advisor rather than a participatory designer or innovator in a development or problem-solving process. To address this issue, ownership in processes and projects, team cohesion, and experiences of relevance are seen as vital conditions for high-level participatory design by several authors (Bowen et al., 2013; Zamenopoulos et al., 2018). As Zamenopoulos et al. (2018) note:

> *"The politics of co-design is also shaped by the view of co-design as a practice that creates more meaningful and relevant futures for people that are engaged in the process, and as a result contributes to their social, economic and environmental sustainability and resilience. The underlying argument is that co-design enables people to take ownership of their environments, services or products and, therefore, creates stronger and more meaningful connections among people and these creations. In community design, this argument has its historical roots in approaching co-design as a tool for protecting local communities from large redevelopment and regeneration plans. In social design, this argument has its roots in approaching co-design as a socially responsible action. In this context, the participation of people in design is a key practice (and mindset) for developing more sustainable and socially responsible futures".*
>
> (Zamenopoulos et al., 2018, 24)

Creating ownership is considered a means in a participatory design process to mitigate inclusion challenges, especially when national policies are to be implemented locally and/or have to be maintained by stakeholders over a longer period. Taking ownership relates to aspects such as stakeholders' interests, knowledge, competence, and decision-making autonomy, and to a certain degree also to control of the project process and its outcomes. Early involvement of stakeholders in a development and/or problem-solving process can contribute to strengthening aspects of ownership. Any inclusion project based on participatory design should initially inform participants about the project phases, their expected contribution, as well as risks and gains. Enhancing autonomy, participants should further understand that they have a choice to join the project. If they cannot choose themselves, they should be given a rationale for why they should. Scepticism about participating can be mitigated at project start, for example by showing successful collaboration cases from earlier projects.

Supportive participatory design includes stakeholders' perspectives and interpretations from the problem definition to the evaluation phase. It encourages discourse about alternatives, which can contribute to the view that the project meets stakeholders' needs and values, and that it will have an impact—this, in turn, increases interest and engagement to participate (Gagne et al., 2005). According to Gagne et al., positive feedback and discursivity in the problem definition and ideation phase also increase competence and responsibility for contributing to successful project accomplishment (Gagne et al., 2005).

Participatory design depends on teamwork, and effective teamwork requires both clear and relevant goals and cohesion (Johnson et al., 2009). Goals motivate behaviour, provide a basis for resolving conflicts, and are prerequisites for the assessment of solutions. Most of the collaborators should feel ownership of the goals, and they should contribute to the success of the group. Cohesion is advanced through inclusion, acceptance, support, and trust, while individuality is endorsed. This is especially relevant when heterogeneous stakeholders and/or marginalised groups or individuals are involved in the participatory design process. An important aspect for successful participatory design within this context is to acknowledge diversity in seemingly homogenous stakeholder groups. This goes for both local communities and marginalised groups, which tend to be seen as having particular characteristics that define them. Reaching an agreement on goals must anticipate that participants have divergent ambitions and motives to join the project. Further, project facilitators must establish different types of communication to coordinate individual perspectives and expectations and to integrate single performances in a common action plan. Participatory design tools that facilitate different types of communication are photos, narratives, or small-scale prototypes produced in and for collaborative workshops (Sanders et al., 2014). Blomkvist and Holmlid (2009) argue that narratives are especially important in the early stages of participatory design processes, as they help illustrate diverse problem perspectives. Bowen claims that sharing stories additionally creates openness, builds trust and ownership, contributes to empathy and cohesion between stakeholders, and alleviates tensions (Bowen et al., 2013). According to Kagan and Duggan (2011), tools such as storytelling and narratives contribute to the flattening of existing hierarchies and levelling of power relationships. They empower participants, increase their visibility, and contribute to confidence building. Equally important, they can provide enjoyable experiences and can emotionally touch and engage people. In settings with marginalised and/or local stakeholders, participatory design of prototypes and blueprints are excellent means for better communication, skills display, and trust building among stakeholders (Keitsch, 2021). Applying narratives, storytelling, boundary objects, and prototype participatory design underlines the socio-cultural dimension of inclusion and emphasises the development of capacities such as literacy, creativity, critical knowledge, empathy, and trust. Trust is built mutually, partly by letting go of the script, engaging in authentic dialogues, and letting the other party take charge. It requires values such as patience with each other and acceptance of diverse views and cultural attitudes, which can be regarded as equally important conditions for trust building as acknowledging others' power, expertise, and professionalism. Social inclusion can, to a certain degree, build on instrumentalised interactions to achieve a common short-term goal, yet long-term partnerships are rarely sustainable without trust and respect for each other (Keitsch, 2021).

Outcomes from participatory approaches and processes range from immaterial results such as giving voice and articulation to citizens' views, emotions and opinions, their feedback, the providing of information, legitimisation, trust building, building of partnerships, knowledge generation, etc., to organisational coherence such as agenda correspondence of different stakeholders, and tangible results such as products, services, buildings, and infrastructure. However, the extent to which participatory approaches achieve desirable outcomes remains open to interpretation, as the literature tends to offer generalised rather than specific or consistent results (Keitsch, 2021).

Allmendinger et al. (2002a, 2002b) and Fung et al. (2003) argue that participatory processes are subject to a range of recurring challenges that can compromise both their effectiveness and the long-term sustainability of their outcomes. A significant concern is the potential for such processes to be corrupted by power imbalances and domination, rendering outcomes vulnerable to manipulation by more influential actors. Additionally, participatory frameworks may neglect or reject alternative decision-making mechanisms such as bargaining or voting that could be equally or even more appropriate depending on the context. Another issue arises when the inclusion of marginalised individuals or groups becomes tokenistic, serving merely as a symbolic gesture to simulate diversity rather than genuinely integrating their perspectives. Moreover, external stakeholders can impose constraints on deliberative decisions and actions, limiting the autonomy of participants. Disengagement is another risk, as stakeholders may lose interest in the process or disagree with its direction or outcomes. Participatory projects and processes can also place unrealistic demands on stakeholders, requiring excessive time, energy, or commitment that is unsustainable. Even when initial results appear successful, they often prove difficult to maintain over the long term, highlighting the importance of designing participatory frameworks that are both inclusive and resilient.

Ethically, empowerment is a core value in participatory design. This notion is, however, not unproblematic and is criticised, among others, for its individualistic, competitive implications and connotations of mastery, power, and control (Riger, 1993). Metaethically, narrative ethics, clinical pragmatism, micro ethics, and empirical ethics[8] are all coherent with PD approaches.

However, sustainable design and participatory design share another significant normative feature: aesthetic participation. Art and design increasingly emphasise the role of user participation and social inclusion, for example by connecting artistic and social activities to co-create and co-design desirable outcomes (Berleant, 2013; Gablik, 2003). The idea of art as "social sculpture" was coined by Beuys in the 1970s (Buchmann, 2006), and since then, aesthetic participation has maintained relevance for including different social groups and cultural practices. As part of cultural practices, aesthetic participation also implies the potential to deliberately change parts of the practice toward sustainability by experiencing and engaging with an aesthetic object, artwork, or natural phenomenon[9]. Elements of aesthetic experience that motivate aesthetic participation can be broadly categorised into three dimensions. First, pleasure-evoking experiences are often tied to consumption, where aesthetic objects hold affirmative value for the subject, offering enjoyment or sensory satisfaction. Second, value-evoking experiences contribute to human development by fostering moral or political awareness, supporting aesthetic education aligned with socio-political goals, or serving as manifestations of prevailing political ideologies. Third, autonomy-evoking experiences arise through encounters with art or natural phenomena that reveal the alien, unfamiliar, or sublime. These moments disrupt conventional routines and trigger new ways of thinking, reflecting personal freedom and promoting self-reflection and transformation (Keitsch et al., 2013).

Participatory design approaches in sustainable public space and placemaking (Ermacora et al., 2016; Hamdi, 2010) often incorporate aesthetic participation as a value-evoking process. This socio-political perspective on aesthetics, oriented toward the pursuit of shared values, can at times imply perfectionist elements where aesthetic experience is equated with accomplishment or even ethical self-realisation. A component of the good life, wellbeing, or just communities is treated here as an essential condition for sustainability. While aesthetic experiences and forms of participation can expand perspectives, deepen emotional awareness, and encourage engagement, they cannot replace ethical judgment or collective political decision-making through democratic deliberation (Singh et al., 2018). The aesthetic-expressive dimension of participation must hence be supported by an emancipatory-normative dimension and, where possible, by an empirical-informative dimension. Designers can play a key role in facilitating the interplay between these dimensions for example by creating spaces and settings that enable inclusive dialogue and collaborative engagement.

Moving beyond material concerns and technical fixes to environmental challenges, contemporary sustainable design concepts increasingly engage with emotional, social, and normative aspects such as product semantics, user inclusion, and design ontologies, which aim to critically address complexity and interdependency of human–nature relationships. The following sections introduce key ideas from Postmodernism as a theoretical foundation for recent concepts in sustainable design, such as critical design, new materialism, and posthuman design, and examine both their transformative potential and their limitations.

3.2.1 Excursion: Postmodernism

According to Swann (2002) the “modernist’[10] development in design subscribes to the positivist canon that objects have real existing primary properties such as size, duration, shape, movement, penetrability, and number, of which an observer can gain objective knowledge. Properties such as warmth, colour, tone, odour, or taste are considered secondary properties, not intrinsic to the object, but as ways in which the object affects someone, e.g., how it appears subjectively.

In 1957, Popper generalised the deductive-nomological method[11] and introduced the “unity of science”, postulating that there is no genuine distinction between propositions and predictions concerning humans, animals, or machines. Sometimes the propositions are complex and can be separated from each other; sometimes this does not work. The point is always to search for correlations between volumes that occur in the physical description of processes, and the method to find the correlations is always the same. Since the deductive-nomological method is valid for all disciplines, a distinction between natural sciences and humanities or social sciences is obsolete. Theories based on this distinction are metaphysical and irrelevant. This means, for example, that Sociology is not placed in the Humanities but becomes part of the unity of sciences as Social Behaviourism.

The universal science concept is both intriguing and repulsive for designers. Some theorists employ Popper's critical rationalism (1996)[12], such as Simon's *Science of the Artificial* (1969); others reject the positivist epistemology. Schön (2016) finds rationalism, for example, too rigorous and abstract to produce professional design, and because it separates means from ends, it separates research from practice, and it separates knowing from doing. By contrast, Schön's reflection-in-action involves looking at experiences, connecting with feelings, and attending to theories in use. A practitioner who frames a situation is allowed to experience surprise, puzzlement, or even confusion in a situation that is uncertain or unique. Reflection-in-action becomes a hermeneutic endeavour, where a designer interprets various interactions with a product (physical, social, and symbolic) as a form of dialogue.

In sustainable design, design for the environment and industrial ecology represent the positivist canon (Opoku et al., 2006), while design for social sustainability and participatory design correspond to hermeneutic and postmodern perspectives that emphasise idiosyncrasy, interpretation and contextual understanding and allow reflections on pluralism, and the questioning of dominant societal ideologies.

The term "Postmodernism" is used by Lyotard (1984), who describes it as a movement characterised by the *"incredulity towards metanarratives"* (1984, ibid., xxiii), while narratives disintegrate into "clouds" of linguistic combinations and language games. Postmodernism repudiates the idea of ultimate scientific, philosophical, or religious truths such as nature, laws, ethics, or God. As Jameson (1991) ironically remarks: *"Postmodernism is what you have when the modernisation process is complete, and nature is gone for good"* (ibid., ix).

Epistemologically, postmodernist philosophy adhering to a social constructivist credo claiming that all understanding of objects is constituted by their historically, socially, and politically influenced linguistic interpretations, and that these interpretations are the only valid representations of the world. Even if discursive interpretation and concrete experience are crucial in postmodernism, and reality only comes into being through human reflections on what the world means, interpretations are necessarily fallible and relative. The goal of a postmodernist discourse is hence not to seek consensus or negotiate meaning (a metanarrative) but to conceive the lack thereof as a productive mechanism for change.

Foucault (1973) uses for example "discourse" as a notion that constructs the topic, defines the objects of knowledge, and directs the understanding of meaning, while simultaneously developing a context of meaning. Discourse controls how ideas are put into practice, what values and norms are related to them, and what kinds of requisites are used to manifest ideas. For example, a discourse on madness may include writings and drawings by people who are considered mad, writings by doctors and administrative people who work with madness and insanity, novels with mad characters, autobiographies of relatives, as well as writing about madness from other disciplines. The requisites in this discourse could be terrifying images of mad people, straightjackets, bars, asylums, hospitals, etc. (Foucault, 1973). Discourses always take place within a historical and socio-political context, and material artefacts such as technology[13] are included in this contextuality, where they both make and warrant information within the discourses, which is often disregarded according to Foucault:

> *"We frequently speak of the technical inventions of the seventeenth century – chemical, metallurgical technology – yet we do not mention the technical invention of this new form of governing man, controlling his multiplicity, utilizing him to the maximum, and improving the products of his labour, of his activities thanks to a system of power which permits controlling them".*
>
> (Foucault, 2007, 146)

Building on Foucault's insight into the mutual constitution of power, knowledge, and material practices, Latour's Actor-Network Theory (ANT) subsequently conceptualises artefacts as integral participants within networks of meaning and power. By positioning non-humans as actants, ANT operationalises Foucault's analysis of the co-implication of material and immaterial dimensions of technology (Latour, 2005). A more detailed examination of ANT is provided in the subchapter "Posthumanism" below at 3.4.

Discourses are connected to power, and for Foucault this relationship is most clearly manifest in forms of domination that seek to regulate, discipline, and control the production and circulation of discourses. Practical subduing via discourse control includes, among other measures, exclusion, e.g., access to products or services, definitions of right and wrong, limitation of contents and participants who are admitted to the discourse, ritualisation of communication, and discrimination against those who are not "normal", "scientific", or "eco-friendly". For example, academic disciplines and political institutions form sustainability discourses through processes of institutionalisation that establish systems of rules and structures designed to preserve, regulate, and disseminate knowledge and information. Notions such as "green growth" make sustainability, for example, more attractive to business, while "pro-environmental behaviour change" aims at social control (Hargreaves, 2010). Reproducing elements of sustainability discourses integrates them into a social domination structure, making them part of commonsensical sustainability.

Postmodernism sees the individual identity of humans as fluid and constantly altered by historical and socio-political contexts. For that reason, the conventional understanding of subjectivity, personality, and the individual human being, synthesising human recognition and existence, becomes problematic. For Foucault, the "subject" itself is a narrative imposed by regimes:

> *"the political, ethical, social, philosophical problem of our days is not to try to liberate the individual from the state and from the state's institutions but to liberate us both from the state and from the type of individualization which is linked to the state. We have to promote new forms of subjectivity through the refusal of this kind of individuality which has been imposed on us for several centuries".*
>
> (Foucault, 1982, 785)

Foucault's deconstruction of "subject" concepts can be compared with Derrida's deconstruction of "word" Derrida (1978) claims that the meaning of a word is unstable; it can have multiple or alternative significances. To explain this, Derrida uses Saussure's (1983) language model which comprises the signifier (a sound or a word: tree, Baum, Arbor, etc.) and the signified (the meaning associated with the signifier: natural being, a tree is a living being). Further, the signification, which describes the relationship between the signifier and the signified, and finally, the sign. which is the outcome of the signification (Chandler, 2019)[14].

Just as identity and recognition do not fully coincide in Foucault's view of the self, the signifier and the signified do not coincide into a fixed meaning in Derrida's theory of language. On the contrary, he argues that the signified can itself take the position of a signifier. For example, the signifier "tree" may relate to the signified "living being", but "living being" can also function as a signifier in a new context, thereby deferring meaning further. Resultatively, the signification procedure ultimately encompasses *"a field of infinite substitutions"* (Derrida, 1978, 25) that are always extendable, reviewable, and ready to be recontextualised.

Derrida's argument illustrates a particular postmodernist critique of positivist language theory, namely that meaning is created through a stable relationship between signifier, signified, and signification culminating in the sign. Conversely, postmodernist language theory claims that the sign is always established through contextual differences, not identity: a "tree" can be signified as a friend or as material for a chair, etc. The meaning of the sign is hence ambiguous, and there is no final relationship between the signifier and the signified, since the signified is always the signifier of another signified. Because of these ambiguities, the real subject of semantics lies in its potential to elucidate: *"not what has been said but what could have been said about an object"* (Foucault, 1988a, 95–98, translation, M.K.). Formally, signs must be part of a language game to have meaning, or as Wittgenstein (1967) puts it: *"A sign lives in the language-game that is its original home"* (ibid., §116). This means that any interaction with signs, production of signs, or attribution of meaning presupposes that the grammar of the language game has to be understood and applied to identify the sign and its way of being a sign. Practically, a sign has to be used repeatedly. The continuous use of signs constitutes part of their stability, even if they can always be reproduced in different contexts, changing their meaning. Saying that a tree is a friend, or that it is useful, relates to different contexts, "friend" to contexts of emotions and relationships, "useful" to instrumental and production contexts.

As convoluted as this excursion may appear in a publication on environmental ethics and sustainable design, it has benefits for concept development in both fields. Postmodernist reflections, when applied to sustainable design, highlight that meaning does not reside inherently in objects themselves. A tree, for example, is not automatically a friend to be protected or simply raw material for making a chair. Instead, meaning is negotiated through various cultural, social, and historical discourses, and it shifts over time. This perspective helps recognise that how nature is valued is not fixed, but shaped by changing contexts and interpretations, which is crucial for developing more thoughtful and adaptable approaches to sustainability. The inherent uncertainty and polyvalence of meanings can encourage the exploration of alternative discourses from diverse cultural viewpoints or user perspectives such as children, thereby fostering the development of innovative, sustainable design concepts.

Before introducing some postmodernist-inspired SD concepts in the following subchapters, it is important to acknowledge some ethical limitations of postmodernist theory, particularly its conceptual indifference to the consequences of discourse. Postmodernist authors tend to focus on how discourses emerge and evolve, while largely ignoring their social and normative implications. As Winner (1993) notes in his critique of postmodern and constructivist approaches:

> *"What the introduction of new artefacts means for people's sense of self, for the texture of human communities, for qualities of everyday living, and for the broader distribution of power in society these are not matters of explicit concern".*
>
> (ibid., 368)

Shusterman (1989) identifies postmodernist theories as a renewal of romanticist ideas and 19th-century aestheticism, which considered art as independent of morality and existing for its own sake.

> *"Decadence and parasitic lack of originality is the very texture of the term postmodern, which takes modernist as its root, adding not a neo or even an anti, which would connote novelty, if only of opposition, but only a post, suggesting mere temporal passage and consequent loss or waning".*
>
> (ibid., 606)

Jameson (1991) criticises postmodernist indifference to capitalist, consumer, and ware ideologies, represented by the belief that there is no depth in any phenomena, which plays right into the hands of capitalism:

> *"postmodernism is not the cultural dominant of a wholly new social order… but only the reflex and the concomitant of yet another systemic modification of capitalism itself – No wonder, then, that shreds of its old avatars – of realism, even, fully as much as of modernism – live on, to be rewrapped in the luxurious trappings of their putative successor".*
>
> (ibid., xii)

Some authors ascribe to the postmodernist relativistic and ambiguous view on morality an inherent value because it responds to a world of increasing uncertainty. For Sterling (2003), a "revisionary mode" of postmodernism, embracing a holistic worldview, is, for example, a suitable guide for sustainability practice. The whole is here not seen as an objective phenomenon, e.g., an ecosystem, but appears as a heterogeneous, contingent interaction pattern of agents (human and non-human) within a complex system that is indefinitely co-created. Gilligan (2012) explains,

> *"This co-evolution … suggests that human beings, through interaction, both make and are made by the environment and are thus actors in their own evolutionary history … This redefines the relationship between humans and the natural environment and recognises reality as a dynamic co-construction, constantly open to revision".*
>
> (Gilligan, 2012, 468)

In conclusion, postmodernism offers valuable supplementary perspectives to environmental ethics and sustainable design. However, its lack of explicit ethical arguments restricts its contribution to epistemological and methodological rather than normative domains. Even if experiences of nature are individual and contingent, they may nonetheless, under certain conditions, reveal collective human needs for engagement with the natural world. This highlights the importance not only of expressive, aesthetic, and semantic sensibilities, but also of an ethical stance opposing the instrumentalisation of the lifeworld and other beings, as expressed among others in critical theory and phenomenology.

3.3 Speculative and Critical Design for Sustainability

> *"There are more ideas on earth than intellectuals imagine. In addition, these ideas are more active, stronger, more resistant, more passionate than politicians think. We have to be there at the birth of ideas, the bursting outward of their force: not in books expressing them, but in events manifesting this force, in struggles carried on around ideas, for or against them. Ideas do not rule the world. But it is because the world has ideas … that it is not passively ruled by those who are its leaders or those who would like to teach, once and for all, what we must think".*
>
> (Foucault, cited in Eribon 1992, 282)

Speculative and Critical Design (SCD) are current examples recognising the postmodernist credo among others via speculative narratives, and the design of critical objects. Most SCD concepts also comprise a normative component, considering, for example, social consequences of design[15]. SCD claims that the utility value of products and services is today surpassed by their semantic and social value and that design practice must be ethically and socially reinterpreted when meanings of products and services change (Marenko, 2015). Design has huge social power, and the designer has the responsibility to critically identify hazardous activities, eliminate them, and develop alternatives. *"Instead of thinking about appearance, user-friendliness or corporate identity, industrial designers could develop design proposals that challenge conventional values"* (Dunne et al., 2001, 59). SCD concepts are, among others, inspired by science fiction and its potential for design innovation and public debates, for example in the form of social dreaming. In sustainable design, social dreaming would, e.g., enable perceptions of "hyperobjects" (Morton, 2013) such as climate change, loss of diversity, and global injustice, which all exceed human comprehension in time and space. Utilising a rhizome approach for comprehensive interpretations, experiences, associations, and dreams are applied in social dreaming to accumulate discrete aspects of hyperobjects. Social dreaming can also be implemented to reverse "defuturing" (Fry, 2009), i.e., design involvement in damaging political, economic, and environmental activities. It enables designer–user dialogues and fosters narratives about preferable societal developments in public debates by employing features of science fiction such as envisioning utopian as well as dystopian scenarios. SCD subscribes to the postmodernist discourse by utilising speculations and conjectures rather than facts and information:

> *"New ideas are exactly what we need today. Conceptual designs are not only ideas but also ideals, and as the moral philosopher Susan Neiman has pointed out, we should measure reality against ideals, not the other way around: 'Ideals are not measured by whether they conform to reality; reality is judged by whether it lives up to ideals. Reason's task is to deny that the claims of experience are final – and to push us to widen the horizon of our experience by providing ideas that experience ought to obey'".*
>
> (Dunne et al., 2013, 12)

Besides ambitious design perceptions: *"We imagined a world where there was a desire for products that met existential needs, reminding us of the frailty of life"* (Dunne et al., 2013, 36), SCD subscribes to the underlying metanarrative that modernity inevitably produces alienation, inequality, and both ecological and human commodification, an assumption surfacing for example in the recurring criticism of the consumer society (Dunne et al., 2013).

Johannessen et al. (2019) describe the initial phase of an SCD process as locating the context and exploring controversial topics such as environmental, social, and ethical issues created by commercial industry, emerging technology, or socio-political structures. The second phase is to ideate to find problems to explore a topic. SCD practice uses a series of "what if" questions (Dunne et al., 2013) as well as speculative future, alternative present, and ad absurdum scenarios where the development of the topic may result in ridiculous, absurd, or impractical situations. The third phase materialises scenarios as narratives, physical or digital objects (props), or a combination of both with the goal to provoke participants to engage with the topic rather than clarifying the topic, and hence SCD scenarios are often open-ended, complicated, and display media and art elements as well as humour and irony. A current example of an SCD approach is Cerpina and Stensli's *Anthropocene Cookbook* (2022) that explores experimental food designs. The rationale of the publication is that the Anthropocene era requires a radical reassessment of food consumption, and consequently, the book displays extreme props, including whisky made from urine and cows that harness energy. SCD-based designs indubitably open new arenas for experimentation and creativity, yet it is not evident whether SCD solutions are more sustainable or better for society than conventional sustainable design practice.

Normatively, SCD contributes to voicing criticism of societal misconducts and encourages debates about preferable futures that might differ from probable futures (Auger, 2013). Design-wise, critical thinking is recommended as the main fundamental for practice, and a designer should question principally everything, in particular attitudes to the commercial industry, technology, and the norms given by society (Mitrović, 2015,).

Ethically, SCD is presented as an alternative to traditional design practices. Dunne et al. (2013) frame it as occupying the moral high ground, implying that SCD designers are responsible for guiding the public toward a "preferable" societal development. This position tacitly assumes that designers possess superior decision-making capacities and moral insight, which qualify them to guide societal stakeholders toward social and environmental well-being. However, the assumption that designers make the "right" choices is neither justified by their training nor by their societal position. Another ethical problem of SCD lies in its rigorist position against capitalism, which may exclude large stakeholder groups from sustainability discourses and hinder the development of inclusive pathways for just transitions and social sustainable entrepreneurship (Keitsch et al., 2013).

Moreover, SCD often lacks critical self-reflection, particularly regarding its roots in white, privileged, middle-class contexts (Ward, 2019). Despite advocating co-creation and speculative futures, it also shows limited engagement with the power dynamics inherent in design, especially the asymmetrical relationships between designers who "empower" or "care for" and the "beneficiaries" of empowerment and care (Akama et al., 2015).

Conclusively, SCD concepts often mirror diverse Western ideologies, values, and methods, and tools, and practices might have limited feasibility for individuals and communities in the Global South. SCD seems, however, beneficial for local sustainable development, while nationally, social and environmental enhancement has to be pursued by all: governments, grassroots movements, and heterogeneous societal stakeholders.

Governments typically pursue tangible targets related for example to the Sustainable Development Goals (SDGs) such as establishing sustainable water, waste, and transport infrastructure, and ensuring access to public services, healthcare, and education. Efforts are implemented empirically, instrumentally, and pragmatically, and should be measurable, even if public participation is incorporated into the process. In contrast, grassroots organisations, NGOs and local groups tend to engage with time- and place-specific challenges and initiatives. Their efforts can benefit from more imaginative and participatory methods, such as social dreaming.

Contrary to postmodernism, which remains indifferent to questions of value and normativity, Speculative Critical Design (SCD) rejects power mechanisms embedded in society and seeks to undermine the conformity and passivity generated by consumerism and commodified aesthetics. Some authors argue that this approach resembles the Critical Theory of the Frankfurt School, even though in a less elaborated form (Bardzell et al., 2013). SCD concepts converge with critical theory in the commitment to resisting the subjugation of the individual and promoting the ideal of collective emancipation. Yet, SCD shaped by postmodernist orientations toward disruption, pluralism, and ambiguity often falls short in cultivating conceptual self-reflexivity. Most notably for this book, its engagement with the human–nature relationship remains immature, nature is interpreted as a backdrop for design activities rather than as a field of ethical and ontological design inquiry. In contrast, critical theory offers profound philosophical engagement with nature experience, emphasising contingency, spontaneity, and non-instrumental value. This perspective can provide an ethical propaedeutics, a preliminary ground for more sustainable design concepts. Accordingly, the following subchapter explores the ideas of Adorno and Horkheimer, whose works remain foundational for understanding the entanglements between the human and natural spheres in modern society.

3.3.1 Excursion: Critical Theory and Aesthetic Experience

The cumulative scientific advancements of the late 18th century not only altered socio-cultural structures but also left philosophy bereft of a teleological understanding of nature. In response, the 19th-century Romanticist philosophy[16] elevated nature to an almost pantheistic status, portraying the artist as a genius whose creations embodied a spiritualised vision of reality. Modernity exalted art as a central model for nature, in which autonomous artworks (the signifiers) represent spirituality (the signified), with both converging into cultural signs that mediate between nature and subjective experience. This turn toward an "inner nature" marks an important stage in the development of autonomous art, as artistic representation increasingly develops into a self-contained sphere for exploring spiritual, existential, and psychological tensions. Rather than depicting nature as an outer environment, artists regard it as a symbolic medium through which the inner life can be projected and examined. Both romanticism and certain strands of modernism read fantasy not as an escape from reality but as a means of constructing and interpreting it, while nature becomes a mirror of the fragmented and troubled self, intertwined with the psychological states of both artist and recipient. Baudelaire's poems, such as *Obsession* (Campbell, 1952), epitomise this tendency: landscapes, forests, oceans, and other phenomena are not mere natural imagery but echoes of inner turmoil, projecting dread, hostility, and longing onto the surrounding environment.

Autonomous artworks surpass the constraints of a reality shaped by industrial modernity and permit both artists and recipients to transcend normality, but they also respond to cultural discontents of modern progress. This response takes concrete form in early 20th-century art movements such as Expressionism and Surrealism, which thematise inner psychological turmoil and critique the dehumanising forces of modernity (Leach, 1999).

Philosophically, criticism of technical rationality, and an all-encompassing industrialisation of society that renders individuals' parts of production and consumption processes, is elaborated upon among others in critical theory (CT). It is difficult to summarise CT concepts, and this excursion reflects mainly ideas of Adorno and Horkheimer from the Frankfurt School's first generation.

Horkheimer and Adorno's critical theory explains socio-cultural structures and phenomena with help of dialectical reasoning.[17] Enabling critical reflection this contributes heuristically (for Horkheimer) to societal liberation by revealing the underlying mechanisms of domination and enabling critical reflection.[18] To benefit society, CT must be explanatory, practical, and normative, i.e., it must pinpoint flaws in current social reality, identify the actors to change it, and formulate both norms for criticism and achievable practical goals for social transformation (Bohman, 2005). However, instrumental, purpose-oriented rationality overshadows critical reflexivity, as well as undermines social human values such as emancipation, equality, and liberty.

Ontologically, CT of the Frankfurt School claims that, for example, mass media and consumer culture are not just regressive to humans but exercise the reification of inner and outer nature, a.k.a. everything they touch: *"...objects become purely thing-like the moment they are retained for us without the continued presence of their other aspects: when something of them has been forgotten"* (Adorno et al., 1999, 321).

CT argues further that mass culture and the culture industry in contemporary societies pacify humans in the wrong way. The easy pleasures which come with material possessions make them tame and indolent. The goal of all industries is to sell their products, and culture industries are factories that produce standardised cultural commodities. The differences between the commodities make them appear diverse, but in fact they are just variations of the same manipulation, and their goal is pseudo-individualisation. Capitalist society objectivates and manipulates the individual by highly restrictive institutions which have little interest in the specific individual.

> *"Life in the late capitalist era is a constant initiation rite. Everyone must show that he wholly identifies himself with the power which is belaboring him ... The eunuch-like voice of the crooner on the radio, the heiress's smooth suitor, who falls into the swimming pool in his dinner jacket, are models for those who must become whatever the system wants. Everyone can be like this omnipotent society; everyone can be happy, if only he will capitulate fully and sacrifice his claim to happiness".*
>
> (Horkheimer et al., 1987, 124)

Epistemically, critical reflexivity remains an opportunity to explore the complexity of human societies, uncover repressive mechanisms and expose weaknesses of social institutions, and eventually contribute to create awareness of, if not liberation from, suppressing structures. Practically, aesthetic experiences of art and nature provide antidotes against the objectivation of human life and instrumentalisation of the lifeworld.

Resistance through aesthetic experience finds an elaborate theoretical articulation in Adorno's *Aesthetic Theory* (1970), which explores art's capacity to negate, resist, and critically reflect the damaged conditions of modern life. Adorno discusses the aesthetic dimension of CT by emphasising the critical function of art in society, while aesthetic experiences and autonomous artworks work as antidotes to both instrumental rationality and the culture industry. Aesthetic experiences are fundamental to humans, and reflexive and productive activities such as art, design, and philosophy are references of these experiences, and their exemplifications are vital for social and cultural development. Aesthetic experiences are also essential for individual growth. The development of an autonomous subjectivity is central to Adorno's philosophy. He argues that what is so alarming in complex, capitalist societies is the extent to which, despite their individualist ideology, these societies in fact ignore and frustrate the individual's experiences of autonomy:

> *"Certainly, art, as a form of knowledge, implies knowledge of reality, and there is no reality that is not social. Thus truth content and social content are mediated, although art's truth content transcends the knowledge of reality as what exists. Art becomes social knowledge by grasping the essence, not by endlessly talking about it, illustrating it, or somehow imitating it. Through its own figuration, art brings the essence into appearance in opposition to its own semblance".*
>
> (Adorno, 1997, 258)

The culture industry endangers art by promoting false desires and meanings, created and satisfied by capitalism. Adorno advocates an avantgarde art because it resists commercialisation and denies the homogenising effects of the culture industry. Instrumental rationality and experimental-scientific interpretation possess a power in society that impedes contemporary art from merely illustrating reality. Modern art therefore focuses on the epiphany of truth in the artwork. The epiphany creates a spiritual relation to the world and promises a fulfilment which can no longer be found in the disenchanted reality. An autonomous artwork lets truth happen and opposes the spotless and synthetic contentment produced by the mass media. Artworks are autonomous when they have no outer purpose and are experienced from within: *"We don't understand music, it understands us"* (Adorno, 1993, 15, translation M.K.).

Adorno explores specifically the role of the "non-identical" of art and nature, in fostering self-reflection and critical reflexivity. The value of the artwork is to serve the non-identical, which is repressed by reality's compulsion to identity. Non-identity means to overcome conceptual imperialism, which impoverishes experience and makes room for new possibilities of thinking. The artwork allows a surpassing experience with phenomena and things, an experience which cannot happen through their instrumentalisation, but through (their) contemplation. By having this potential, artworks illuminate the "non-identical" and show possibilities to alter one's lifestyle and simultaneously provide a safe and playful possibility to explore new thoughts. Adorno's concentration on artworks signifies a return to the subject and emphasises the value of the concrete aesthetic experience for humans. The non-identity of the world, mediated through an artwork, encourages the subject to become autonomous, self-reliant, and resistant against the manipulations of the culture industry.

The social value of autonomous art is then not just to overcome the limitations of a petrified ego but also to realise the functionalist society as a *"caricature of true life"* (Adorno, 2005, 15). Art can halt the self-destructive process of purpose rationality, instrumentalisation, and aesthetic mass-consumerism and symbolise reconciliation by making the repressed and the locked-out visible.

> *"At ground zero … where Beckett's plays unfold like forces in infinitesimal physics, a second world of images springs forth, both sad and rich, the concentrate of historical experiences that otherwise, in their immediacy, fail to articulate the essential: the evisceration of subject and reality. This shabby, damaged world of images is the negative imprint of the administered world".*
>
> (Adorno, 1997, 31)

Art can embody the "inner nature' of humans and symbolise reconciliation between the natural and the human-made world, and the aesthetic experience is able to approach the non-identical without the imperialism of explanation. Consequently, Adorno puts his hope for a reconciliation of inner and outer nature in the aesthetic revelation that both an autonomous artwork and natural phenomena can create, advancing them to representations for a non-manipulated reality.

> *"Natural beauty is the trace of the nonidentical in things under the spell of universal identity. As long as this spell prevails, the nonidentical has no*

> *positive existence. Therefore, natural beauty remains as dispersed and uncertain as what it promises, that which surpasses all human immanence. The pain in the face of beauty, nowhere more visceral than in the experience of nature, is as much the longing for what beauty promises but never unveils as it is suffering at the inadequacy of the appearance, which fails beauty while wanting to make itself like it. This pain reappears in the relation to artworks".*
>
> (Adorno, 1997, 73)

Adorno does not explicitly develop an ethics concept, but authors such as Nelson (2011) and Reno (2023) argue that his philosophy lays the groundwork for an environmental ethics, based on non-instrumental, reflective engagement with the natural world that resists both domination and idealisation.

Implications of Adorno's theory for sustainable design are, first, that interpreting artworks' connections and contradictions to societal conditions trains critical reflexivity. Dieleman (2008) mentions four forms of reflexivity related to artworks' significance for sustainability: aesthetic reflexivity expressing symbolic meaning, hermeneutic reflexivity facilitating reflections on everyday routines and disruptions where artworks can play important roles as agents for social change, ontological reflexivity which allows one to perceive variations and deviations of reality, which artworks can initiate, and professional reflexivity, i.e., linking knowledge, emotions, and experiences for developing contextual understanding, where artworks can enhance abilities to experience surprise, puzzlement, or confusion, and open space for reflection through practising non-identification, perceiving the entire differentness of the other.

Second, in relation to ethics, artworks connect inner and outer nature, and even if real reconciliation is not achievable, art teaches one to strive for it anyway. In terms of environmental ethics, CT would classify as anthropocentric, regarding human normativity as developed through everyday life and its constituting structures by society. Some of CT's ideas on the human–nature relationship can be read as a prologue to posthuman design. CT criticises, for example, the assumed moral superiority of humans as a means for dominance over nature (Adorno, 1993) and emphasises the mutuality of human domination of each other and of nature (Horkheimer et al., 1987):

> *"What human beings seek to learn from nature is how to use it to dominate wholly both it and human beings. Nothing else counts. Ruthless toward itself, the Enlightenment has eradicated the last remnant of its own self-awareness".*
>
> (Horkheimer et al., 1987, 3)

Expanding the critical potential of aesthetic experience, Adorno's notion of the non-identical builds a philosophical bridge between the human and non-human, challenging instrumental reason and moderating nature domination through values of mutuality, respect, and empathy. Non-identity breaks the domination of inner and outer nature via (aesthetic) experiences, carrying an indeterminate anticipation of peace, or as Adorno puts it:

> *"The hope excited by the presence of what is most ancient, is that animal creation might survive the injustice done to them by human beings, if not humanity itself, and bring forth a better species, which finally succeeds … the hope that animal creation might survive the wrong human beings have done it".*
>
> (2005, 115)

Conclusively, critical theory's human–nature discourse challenges anthropocentrism and supplements ethical rationality and discursivity with interpretations of vital phenomenological experiences of inner and outer nature, while its conceptual groundwork outlines a theoretical trajectory toward post-anthropocentric debates in posthumanism and new materialism.

3.4 Posthumanism

Criticism of rationalist philosophy peaked at the turn of the 21st century, giving rise to new epistemological and ethical nature interpretations, most notably posthumanism and new materialism[19] which offer alternative frameworks for reimagining the human–nature relationship. Nature is reimagined here not merely as anthropocentric backdrop but as an active participant in the human cultivation of empathy, ethical responsibility, and aesthetic or existential liberation. Authors such as Latour (2005), Haraway (2007), and Braidotti (2016) emphasise the mutual entanglement of human and nonhuman agencies, placing relationality and co-embeddedness of all beings at the core of their concepts.

Humans are no longer perceived as central agent but instead radically recontextualised or even rendered marginal within the broader dynamics of life, matter, and evolution. Decentring the human, posthumanist concepts critically engage with the illusion of anthropocentric superiority, drawing, for example, on Nietzschean philosophy to exemplify what Lyotard (1997) called the *hybris of reason* i.e., the excessive confidence in human rationality and its presumed centrality within the cosmos. According to Iveson (2015) Nietzsche's fable of the "clever animals" exemplifies Nietzsche's posthumanist critique of anthropocentric metaphysics by portraying human knowledge as a fleeting and inflated construction within an indifferent cosmos:

> *"In some remote corner of the universe, poured out and glittering in innumerable solar systems, there once was a star on which clever animals invented knowledge. That was the highest and most mendacious minute of 'world history' – yet only a minute. After nature had drawn a few breaths, the star grew cold, and the clever animals had to die. One might invent such a fable and still not have illustrated sufficiently how wretched, how*

> *shadowy, and flighty, how aimless and arbitrary, the human intellect appears in nature. There have been eternities when it did not exist; and when it is done for again, nothing will have happened. … There is nothing in nature so despicable or insignificant that it cannot immediately be blown up like a bag by a slight breath of this power of knowledge; and just as every porter wants an admirer, the proudest human being, the philosopher, thinks that he sees on the eyes of the universe telescopically focused from all sides on his actions and thoughts".*
>
> (Nietzsche, 1873,1)

According to Herbrechter (2013), posthumanism appeared in the late 1950s as an offspring of, among others, postmodernist and poststructuralist theories, and evolved into various lines of thought, the first phase starting with poststructuralist critiques in the 1980s, followed by a phase of discussions on what it means to be human, changes of subjectivity, body enhancement, cyborgs, and novel interspecies relationships, for example in Haraway (2016a): *"The cyborg appears in myth precisely where the boundary between human and animal is transgressed. Far from signaling a walling off of people from other living beings, cyborgs signal disturbingly and pleasurably tight coupling"* (2016, 7). The third ongoing phase, the ecological "posthuman turn", is characterised by, among others, new materialism and object-oriented ontology (see also subchapter 3.5 on new materialism).

Epistemologically, posthumanists adopt the "speculative turn",[20] seeing humanist philosophy as incapable of dealing with current challenges such as the ecological crisis, cybernetic risks, and biopolitical issues. Speculative philosophy seeks to access the "real world" beyond socially and linguistically constructed structures, but it remains bound to human existence as its default point of reference (Rekret, 2016). The tension between overcoming correlationism and reproducing anthropocentric thinking is among others evident in Meillassoux's (2010) concept of the "ancestral", which denotes a reality that exists and is knowable independently of human thought. However, the term still depends on human rationality, for example, by presuming that the mind is capable of understanding nonhuman or pre-human realities and the difference between them.

The paradox of requiring human abstraction to conceptualise the nonhuman also extends to posthumanist theory. Posthumanism examines complex entanglements between human, nonhuman, and technological domains. However, critics have pointed out that posthumanism epistemologies often preserve the anthropocentric thinking structures they seek to dismantle, among others by relying on theories that privilege abstraction (Rekret 2016). Herbrechter (2013) finds that some posthumanist concepts resemble Foucauldian discourse, especially when applied to concrete socio-political contexts such as the Human Genome Project, biotechnology, or surveillance regimes. In such cases, the focus often shifts away from the agency of the nonhuman and back to how humans interpret, regulate, or narrate these processes, thereby confirming human centrality in ways that contradict posthumanism's deconstructive aims. A convincing example for this dilemma appears in Haraway's (1997) OncoMouse™, which offers a critical account of the commodification and instrumentalisation of nonhuman animals within scientific and capitalist systems. Yet, calling a mouse a "modest witness" reminds us of anthropomorphism, rather than illustrating the mouse as a nonhuman agent with its own ontological standing.

It appears that posthumanist approaches necessarily rely on anthropocentric epistemologies, raising questions about the feasibility of genuinely non-anthropocentric theory. This shortcoming also calls for the exploration of alternative human–nature relationships such as aesthetic experience and interaction which may offer more embodied, affective, and relational ways of encountering the more-than-human world, as further discussed in Chapter 5.3 of this book.

In terms of ethics, human acceptance as being part of the material world is seen as a benchmark for an appropriate moral and political standpoint. Some posthumanist authors justify this argument with a teleological perspective. Sharp (2011) and Braidotti (2016) refer, for example, to Spinoza's pantheist philosophy to correlate the divine, nature, and humans through a vitalist telos (see also Bennett, 2010). Spinoza's Neoplatonic cosmotheology[21] postulates that the world is a coherent whole because God and nature are one, and that knowledge (both in the form of contemplation and understanding) of nature corresponds to knowledge of God: *"Whatsoever exists expresses God's nature or essence in a given conditioned manner … that is, … whatsoever exists, expresses in a given conditioned manner God's power, which is the cause of all things, … therefore an effect must necessarily follow"*. (Spinoza, 2009, Proof, prop. XXXVI).

The intrinsic relationship between God, nature, and humans establishes a foundation for recognising and following a divine natural order, which in turn offers guiding principles for human moral action. Sharp (2011), through a feminist reading of Spinoza's vitalist philosophy, explores how values, intentions, and actions are deeply entangled with natural phenomena. This results in a pragmatic maxim:

> *"Action becomes an endeavor to cultivate a sensuous receptivity, in order better to determine the relations of composition that most enable one to think and thrive. Since one can exist only by virtue of the affects that circulate in one's environment, etiological* [22] *ethics entails the development of mutually beneficial affective compositions".*
>
> (2011, 216)

From a negative dialectics perspective, emotional entanglement becomes particularly apparent in occurrences that exceed human control, such as ecological disasters and the enduring trauma they produce. The term "eco-haunting" captures the unsettling consequences of environmental harm, emerging through images and experiences that evoke past or ongoing emotional responses (Barad, 2010). Konior (2021) elaborates on this through an analysis of representations of dead or mutated marine life in post-Fukushima Japanese art, interpreting them as manifestations of eco-haunting spectral reminders of nuclear contamination and ecological crisis that rupture both ecosystems and collective memory. In summary, posthumanist ideas of affective and material entanglements underscore the intrinsic interconnectedness of human actions and natural phenomena. From an epistemological standpoint, acknowledging action–nature dependency supports the integration of cognitive reflexivity with emotional and aesthetic awareness, thereby enabling more embodied, affective, and situated ways of knowing (Dieleman, 2008).

Anti-subjectivity, established on a universally assumed humanist subject–object polarity, is a core ontological feature of posthumanism. Several theories build on Actor-Network Theory (ANT), represented by e.g., Latour (2005), who sees science as an attempt at truth construction by human actors. Scientific objects have no other truth than the facts that are established by scientists and the efforts that they make to prove them as scientific data (Kipnis, 2015). ANT aims to remove boundaries between nature, society, and technology and rejects any subject–object relationship, perceiving both humans and non-humans (here primarily technology) as equal actors. All acting entities (actors), including people, technology, and non-living or inanimate objects, exist in a network that comprises undefined relationships between them. However, all actors want to achieve a goal and hence recruit collaborators to follow the same goal, thereby stabilising and increasing the network and ergo their power.

ANT assumes that social relationships are built into technologies, shaped, and controlled by the network and its main actors. The industrialisation of products is, for example, influenced by mass production technology (actor) as well as capitalism (actor) or communism (actor). This presumes that society and technology co-evolve. ANT assumes, however, that the consequences of society are built into technology, which limits the socio-political power of specific actors. Critics of ANT claim, hence, contrarily that social (cultural, ecological, and ideological) circumstances determine the use and significance of a technology. In the example above, capitalism and communism implemented mass production for different purposes, while production and product design represented ideological messages of the respective socio-political structure. From a pragmatic perspective, ANT is criticised because the concept does not allow predicting the future of an actor or a network, which makes it difficult to include potentials, hopes, and expectations for possible futures (Harman, 2009).

A connection between ANT and posthumanism lies in the latter's appeal to humans to dissolve human and nature, subject and object, organic and inorganic antinomies, and accept the intrinsic vitality of materials and their role as "actants" (Bennett, 2010). Bennett characterises actants by their capacity to "animate, to act, to produce effects dramatic and subtle" (ibid., 6) and claims that humans are composed of a complex web of active actants as well.

> *"Vital materialism better captures an 'alien' quality of our own flesh, and in so doing reminds humans of the very radical character of the (fractious) kinship between the human and the nonhuman. My 'own' body is material, and yet this vital materiality is not fully or exclusively human. My flesh is populated and constituted by different swarms of foreigners … the bacteria in the human microbiome collectively possess at least 100 times as many genes as the mere 20,000 or so in the human genome … we are, rather, an array of bodies, many different kinds of them in a nested set of microbiomes".*
>
> (ibid., 112–113)

Bennett correctly points out that matter cannot be fully understood through positivist attributes or nomothetic deduction alone, but must also be encountered affectively, through its capacity to act upon and resonate with environments, bodies and sensibilities. However, as Rekret (2016) argues, the attempt to address non-identity by positioning humans as passive recipients to nature risks oversimplifying the complex relation between the mental and the material and this becomes particularly problematic when ethical responsibility is merely grounded in subjective affective response, while overlooking broader social relations and historical developments that shape material experience and agency.

Ontologically, posthumanism sees humanity as a transient event in the evolution of complex life forms dismissing humans' phylogenetic extraordinariness (Herbrechter, 2013). Ethically, anthropocentrism is rejected for its tendency to present the exploitation of nature as natural or self-evident. In response, Loh (2022) proposes an "Inclusive Ethics", which prioritises relationships over isolated entities or fixed categories, placing agency, responsibility, and justice at the center of ethical inquiry. Loh further expands this to include specific relational contexts, such as the ethics of human–machine interactions, thereby underscoring the need for a flexible and inclusive posthuman ethics.

Haraway's "Ethics of Kinship" and MacCormack's "Vitalistic Ethics" support ideas of the profound interconnectedness of human, nonhuman, and technological life. Haraway suggests connecting with non-human beings and moving away from narratives of the Anthropocene, where humans are the protagonists of history, to the Chthulucene, which emphasises "chthonic"[23] earth-bound interpretations, and creates narratives about relationships and interactions of many creatures: *"Unlike either the Anthropocene or the Capitalocene, the Chthulucene is made up of ongoing multispecies stories and practices of becoming-with in times that remain at stake, in precarious times, in which the world is not finished and the sky has not fallen – yet"* (Haraway, 2016b, 55). Haraway essentialises the interrelatedness of beings and environments, however accepting that agency between them is not always symmetrical since humans have, e.g., power over other beings. She then develops the idea of response-ability, which postulates that instead of presupposing and imposing ethical frameworks for interaction with the world, humans should consider and act according to their entanglement with other life forms, being affected by co-existence and developing an atunement to the needs of other beings (Hofman, 2023)[24]. Similarly, MacCormack (2012) founds her ethics on the entanglement, here of human and non-human bodies. Ethics only becomes problematic when nonhuman actors are linguistically interpreted and compared to human actors, because both semantic interpretation and comparison signify anthropocentric colonisation and epistemic violence and are hence unethical:

> *"Posthuman Ethics has consistently sought the silencing of what is understood as human speech emergent through logic, power, and signification. Human speech makes the world according to the human, tells the world what it is and speaks for the world, that is, to other humans and to the gods of human speech – religion, science, capital. Silencing human speech opens a harmonious cacophony of polyvocalities imperceptible to human understanding, just as human speech has the detrimental effect of silencing unheard, unthought expression".*
>
> (2012, 144)

When the non-physical (abstract) is dismissed, non-transcendental (physical, sensual, and experience-based) relations to the world emerge, and a vitalist philosophy can replace anthropocentric nominalist and linguistic doctrines. The agents in MacCormack's vitalist philosophy are immanent external life forces to relate to without comprehension, while this type of relating allows the development of liminal[25] thinking, thereby transforming perceptual human limitations. Most posthuman ethics criticise conditions of conventional ethics such as rationalism, language, and human autonomy, and aim to suspend the morally privileged position of humans. The focus on a novel understanding of materiality, which is also discussed in the next chapters, is employed in some concepts to transgress intellectual boundaries, to augment traditional anthropocentric and biocentric views.

Ethical shortcomings of posthumanism relate, among others to the potential instrumentalisation of concepts by biologism and biopolitics (Radomska, 2016). The surpassing value of organic life over the life of humans and biologistic metaphors might also result in ostracism and discrimination against single forms of life. There is also a latent danger of reductionism when beings are degraded to mere components of nature. Finally, ANT's pragmatic agent-action focus underestimates imaginative and contemplative moments "when species meet' (Haraway, 2007) or dismisses them as "transcendental". In terms of moral decision-making, posthumanism assigns rights to all (living) beings, but eventually human beings have the duty to maintain these rights. Soper (2012) suggests preserving human responsibility, for example when addressing environmental challenges because posthuman concepts always presume a human community as addressee for acknowledging non-human life. Sundberg (2014) criticises the posthumanist narrow-mindedness in simplifying Enlightenment doctrines and stereotypes of the nature and culture divide as universally valid:

> *"As a consequence, Anglo-Eurocentric concepts retain their place as the only ones that matter in posthumanist thought (ibid., 40), thereby ignoring indigenous epistemic traditions: …'more-than-human methodologies currently elaborated in posthumanist geographies are but one approach in a world of inclusive or non-dualist frameworks'".*
>
> (ibid., 36)

Counteracting colonising posthumanist predispositions, Sundberg suggests "walking with" as phenomenological and bodily engagement with the environment:

> *"I posit walking as key to decolonizing in order to highlight the importance of taking steps – moving, engaging, reflecting – to enact decolonizing practices, understanding that decolonisation is something to be aspired to and enacted rather than a state of being that may be claimed".*
>
> (ibid., 39)

The "walking" approach is also applied by sustainable design and discussed below.

In my view, posthumanism emerges as **a** response to the dire consequences of decades of ecological over-exploitation, culminating in a series of escalating environmental crises. These crises also expose the limitations of human-centred paradigms, leaving many with an intensifying sense of disorientation, vulnerability, and helplessness. Science, while offering technical solutions, often finds itself trapped in a vicious cycle developing increasingly efficient eco-technologies that remain structurally difficult to implement, especially when such solutions conflict with economic imperatives or fail to align with profit-driven models. Academic knowledge is governed by management principles in terms of standardisation, efficiency, and quantitative measurability, leaving little time for critical reflections. Collective and individual ethical decisions often appear powerless or insignificant in the light of global neoliberalism. Against this backdrop posthumanism offers epistemological, methodological, and ethical alternatives. Braidotti (2013) states: *"The posthuman condition urges us to rethink the human in ways that disengage it from the classical legacy of humanism and its presumed universalism, and to develop new normative frameworks based on an ethics of sustainability, responsibility, and interconnection"* (ibid., 49).

Regarding human–nature relationships, posthumanist concepts encompass a range of perspectives: some frame nature as a moral or spiritual ideal; others, influenced by critical theory, emphasise the non-identity of nature and its construction as "the other"; while still others draw on Foucault's (1984) notion of *heterotopias*, real, layered spaces that simultaneously mirror and subvert dominant cultural orders (Clarke and Rossini, 2017).

Ethical maxims of posthumanism comprise awareness, atunement, empathy, engagement, care, and responsibility for the world. Posthumanist concepts hold a high potential for motivating moral actions rather than merely fostering moral reasoning. Some concepts appear as "postmodernified" interpretations of biocentrism, ecocentrism, or physiocentrism, and hence face similar objections. Yet, despite certain argumentative shortcomings, posthumanism has a considerable potential to diminish the nature–culture gap henceforth.

This entails a set of methodological and ethical regulative[26] principles that a relevant for contemporary environmental ethics and sustainable design.

- First, it presupposes perceiving nature within real-life cultural and environmental contexts, recognising that such perceptions are always situated and mediated by local practices, histories, and ecological conditions.
- Second, it implies the commitment to exploring nature and culture not as discrete or opposing categories but as intrinsically entangled and co-constitutive, thereby challenging enduring dualisms that dominate Western epistemologies.
- Third, it necessitates acknowledging aesthetics, sensitivity, and emotions as legitimate and indispensable sources for understanding the complex interrelations between beings, positioning these affective and experiential dimensions as equally significant alongside scientific analysis and moral reasoning.
- Fourth, it requires communicating with nature, an endeavour including the readiness to engage with forms of meaning, agency, and expression that may exceed anthropocentric understanding.
- Fifth, it calls for the cultivation of liminal contemplation, a reflective openness to thresholds of experience where conventional boundaries between subject and object, self and other, human and nonhuman, may become productively blurred.
- Sixth, it suggests co-acting in situ with other beings, emphasising direct, embodied participation within shared spaces of existence as a means of fostering relational ethics and mutual responsiveness.

In the context of this book, the principles function not only to inform the development of ethical theory but also to guide practical implementation. In doing so, they bridge abstract insights and concrete socio-ecological applications, ensuring that sustainable design is approached as both a conceptual framework and a lived, material practice.

Before turning to phenomenology, which can complement traditional environmental ethics and enrich approaches to sustainable design, it is relevant to briefly consider the role of posthumanism in design thinking beyond anthropocentric boundaries.

Authors such as Gatto et al., 2019; Pettersen et al., 2018; and Rice, 2018, recently integrated posthumanist ideas in design projects for example as method that encourages designers to learn how to interact in multispecies settings. Gatto et al. (2019) view design as "process-oriented critical instrument" (ibid., 2) and start their speculative design approach with acknowledging other beings, attempting to make their voices heard, and eventually co-imagining heterotopias as alternatives to current structures. Practically, Gatto et al. (2019) worked with awareness creation for soil contamination. Their installation "Geomerce" is based on a project, which *"investigates how the physiological behaviours of endemic flora can provide means to geographically situate issues of soil pollution and imagine alternative agricultural practices to be developed on existing geographical areas"* (ibid., 4). Incorporating "learning from nature" is frequently applied in design among others in biomimicry; however, the ethnographic concept of "following the plant" is novel. Plants are recently discussed in ethnography as an *"opportunity to test the strictures of anthropocentrism and the possibility of reorienting the place-based attention to sociality"* (Hartigan, 2019, 1). Gatto et al.'s (2019) *"multispecies methodology"* (ibid., 7) merging scientific, sociological, ethnographic, and aesthetic methods, displays a liminal designerly interpretation of posthumanism, and practical findings of the study document for example that relational narratives can visualise non-human positions, thereby contributing to enhance designers′ agency and the willingness to pursue all-encompassing solutions. Pettersen et al. (2018) address climate change and sustainable urban development via the posthuman approach of including trees. The goal is to front urban ecosystems by *"the participation of trees in design processes … we propose to think of agency in terms of the trees' ability to produce effects"* (ibid., 2). Trees are considered regarding their historic and geographic significance (situatedness), and similar to human participants, they convey cultural heritage, but they are also regarded as reflectors of biodiversity, soil and air quality, and climate. Pettersen et al. point rightly to the fact that ecosystems cannot participate in a discourse be themselves and so designers have the task *"to find ways of inviting them"* (ibid., 4).

In summary, posthuman ideas facilitate perceiving and visualising interdependencies between humans and nature, heuristically assuming an intrinsic connection between both. Another advantage of posthumanism for environmental ethics and sustainable design, which is discussed more detailed in Chapter 5.2, lies in supplementing rationalist knowledge with e.g. idiosyncratic knowledge[27] perceiving humans and non-humans within similar contexts, thereby expanding reflection and agency spaces. Yet, it is also necessary to recognise that posthumanism concepts often operate on the basis of certain contestable hypotheses, such as the attribution of agency or actor-status to artefacts (e.g., Rice, 2018)[28]. Considering, for instance, that the relationships between artefacts, their production, use, and disposal are human-defined, the notion of artefact agency is a questionable claim that warrants critical reflection of posthumanist concepts rather than merely emulating them for a desired outcome.

3.4.1 Excursion: Phenomenology

The following excursion introduces the phenomenological philosophy of Merlau-Ponty, and, more elongated, Heidegger's thinking. The aim is, first, to highlight the intersections between posthumanism and phenomenology, particularly their shared emphasis on the interdependence of corporeality, sensory experience, and nature as co-constitutive source of meaning and, second, to introduce some ideas later reformulated within new materialist concepts.

Merleau-Ponty's phenomenology sees corporeity and corporeality (the state of being corporeal) not as mere conditions to connect existences within the world but as essential for being-in-the-world. Sensual perceptions are grounded in the body's engagement with the world and are essential to constituting a relationship between the experiencing and the experienced. By relocating the sensitive and sensible subject in an environment with which it interacts, Merleau-Ponty (2002) establishes the human–nature relationship as an interactive performance of an individual that simultaneously perceives itself and the affection of this self by the environment. Perception plays here a double role, as a fundamental element of self-insurance and self-discovery and as a mediator of "the other" through sensory impressions. Far from being merely a basic combination of physical and mental processes, as in behavioural philosophy, perception is equivalent to being-in-the-world; that is, to inhabiting the lifeworld[29].

Thinking is grounded in pre-reflective sensory consciousness, and rational judgements always imply instinctive understandings that serve as their foundation, such as the organisation of one's own body as something that can be experienced as a systematically structured organon corresponding with a similarly structured external lifeworld.

> *"The theory of the body schema is, implicitly, a theory of perception. We have relearned to feel our body; we have found underneath the objective and detached knowledge of the body that other knowledge which we have of it in virtue of its always being with us and of the fact that we are our body. In the same way, we shall need to reawaken our experience of the world as it appears to us insofar as we are in the world through our body, and insofar as we perceive the world with our body. But by thus remaking contact with the body and with the world, we shall also rediscover ourselves, since, perceiving as we do with our body, the body is a natural self and, as it were, the subject of perception".*
>
> (Merleau-Ponty (2002), 239)

Merleau-Ponty distinguishes the lived body from the physical body, the former signifying an active and engaged entity with ontological relevance because of its abilities to inaugurate relations between the subject and an object and to transcend both, for example by "becoming part of the surroundings" (Halák, 2021). The dissolution of boundaries between subject and object sometimes to the point of rendering the subject insignificant is an important feature of posthumanist thought.

Human–nature interdependence manifests in the fact that the individual body exists in a particular natural and socio-cultural environment, a life-world, both constitutive as an onset for distinguishing between the self and that very environment. The dialectics of world and experiences of the self are represented by reflexivity, which is, however, not self-referential or solipsistic but finds its counterpart, or supplement, in the outside world (Sousa, 2018). For instance, the body expresses its intentionality through its permeability to the world and its beings, i.e., perception of and interactions with things and beings, or as Merleau-Ponty (2002) puts it: *"The body's motion can play a part in the perception of the world only if it is itself an original intentionality, a manner of relating itself to the distinct object of knowledge. The world around us must be, not a system of objects which we synthesize, but a totality of things, open to us, towards which we project ourselves"* (ibid., 450). For Merleau-Ponty, a pre-reflective sensory consciousness harmonises objective and subjective distinctions of human experience. For example, rational judgments presuppose instinctive understandings of one's own body as something that exists in a specific surrounding, a systematically structured organon within a similarly structured external world, experienced somatologically as *"preconscious unity with the environment"* (Berman, 1989, 25).

Both Merleau-Ponty and Heidegger credit the ability to generalise experiences and hence exercise rational reasoning, but they are also aware that the same rationality removes humans from reality and from experiencing an immediate meaningfulness of the world. While Merleau-Ponty assumes, however, the interdependency of human corporeality and a life-world as prerequisites for objects to be(come) meaningful, Heidegger (1996) insists on the inherent values of objects for human ways of being (Dasein). Objects have value as "ready-to-hand" (Vorhandenes) instead of existing only for instrumental purposes "present-at-hand" (Zuhandenes). Before discussing how shifts from "ready-to-hand" to "present-at-hand" are initiated either by a so-called "breakdown" or by ontological integration via artworks and discussing their implications for sustainable design, a brief note on Heidegger's ontology.

Heidegger's *Being and Time* applies existential phenomenology[30] and hermeneutics to investigate the meaning of being (Sein)[31]. The human being is called Dasein (Being-there). Dasein signifies an individual, which is fundamentally concerned about its own way of being. It is (to a certain degree) conscious of its body and mind, reflects on its existence, and intends to exist in a certain way. Since Dasein is so engaged in contemplating and transforming its existence, its main feature is "care" (Sorge). This implies taking care of oneself, but also of other beings, nature, and things. "***The essence of Dasein lies in its existence***" (emphasis Heidegger).

> *"The characteristics to be found in this being are thus not objectively present 'attributes' of an objectively present being which has such and such an 'outward appearance' but rather possible ways for it to be and only this. That that-ness of this being is primarily being. Thus, the term Da-sein which we use to designate this being does not express its what, as in the case of table, house, tree, but being".*
>
> Heidegger,1996, 40)

The genuine constitution of Dasein is *"being-in-the-world"*, which means that individuals are engaged with the world through purposeful dwelling, not opposed to it, as objective observers of various phenomena. Being-in-the-world as familiarity with the world is the first existential mode of Dasein. The first route to experience subject-world coherence is via the relationship between Dasein and other beings, for example useful things, which Heidegger calls "equipment" (German "Zeug", used synonymously with "objects" and "things"). Equipment comprises, for example, devices for writing, sewing, working, transportation, measurement, etc. The equipment always stands in a co-constitutive, reversible, and dynamic context with its creator and its user; for Heidegger, there are no things without being involved with them. *"We shall call the useful things kind of being in which it reveals itself by itself **handiness**"* (emphasis Heidegger).

"It is only because useful things have this being-in-themselves, and do not merely occur, that they are handy in the broadest sense and are at our disposal" (1996, 65). … *The work produced refers not only to the what-for of its usability and the whereof of which it consists. The simple conditions of craft contain a reference to the wearer and the user at the same time. The work is cut to his figure; he 'is' there as the work emerges. This constitutive reference is by no means lacking when wares are produced by the dozen; it is only undefined, pointing to the random and the average. Thus, not only beings which are at hand are encountered in the work but also beings with the kind of being of Dasein for whom what is produced becomes handy in its taking care. Here the world is encountered in which wearers and users live, a world which is at the same time our world"* (1996, 66).

Heidegger argues that the central existential problem lies not in humanity's relation to or use of equipment *per se*, but in the growing fixation on the scientific and technological dimensions of the world; what he refers to as its *equipmental* character. This preoccupation fosters an increasingly instrumental attitude that prevents deeper reflection on the question of human existence or *Being* itself.

In response, Heidegger calls for a renewed engagement with this fundamental ontological question. Ontologically, human beings are always situated in a pre-scientific and pre-linguistic context, constituted through their everyday interactions with objects understood here to include both artefacts and elements of the natural world. This means that the subject is not separated from the world of external objects but lives in specific physical, cultural, and historical surroundings of which objects are part, that subjects and objects "share" environments, so to speak. Human action takes place in a world of things (Zeug) that are part of a concrete world (lifeworld) and can be used; they are ready-to-hand (zuhanden, Zuhandenes). All things are useful and, therefore, usefulness is the essence of things or equipment, it is their being. In that way, a pair of scissors is ready-to-hand as a cutting tool, a chair as a sitting instrument, a house as a living place, a park as a recreation environment. Humans experience things within their readiness-to-hand. Readiness-to-hand means that, in daily practice, the particular thing is not explicitly perceived as a thing, it is just there, ready to use. However, things can change their character and attributes and become something other than useful, and then they turn from ready-to-hand to present-at-hand (vorhanden, Vorhandenes).

Heidegger's term "breakdown" signifies the act of things changing, for example by being destroyed or failing in one way or another. When things are missing their taken-for-granted functionality, they start to show different attributes, allowing humans to experience their presence-at-hand. A breakdown followed by a presence-at-hand experience might trigger astonishment and shock; it can evoke reflections on the nature of things and motivate novel scientific, material, aesthetic, moral, and instrumental perceptions and actions. Art can initiate a breakdown deliberately to alter the readiness-to-hand perception of things, as, e.g., Beuys' *Fat Chair* (Tate, 2025). The "breakdown" concept is a vital component in Heidegger's reversal of the traditional subject–object antinomy that believes in a detached subject, which contemplates or instrumentalises the rest of the world. Instead, he claims that humans experience themselves (their Dasein, i.e., existential nature) through practical interaction with things, and sometimes, through the breakdown as forced change of this interaction. The dialectical process by which things transition from readiness-to-hand (Zuhandenheit) to presence-at-hand (Vorhandenheit) through breakdown (Ward, 2020) holds particular significance for examining aesthetic experience and advancing sustainable design practices, as it diminishes anthropocentric dimensions by allowing things to appear from a non-instrumental perspective.

The second route to experience subject–world coherence is art, which Heidegger also sees as a mode of being. The meaning of art is to reveal truth (aletheia). This is programmatically articulated in the text *The Origin of the Work of Art: "The artwork opens up, in its own way, the being of beings. This opening up, i.e., unconcealing, i.e., the truth of beings, happens in the work. In the artwork, the truth of beings has set itself to the work"* (2002, 19). That art reveals truth can, however, not be interpreted as if all production (*poesis*) equals truth revelation. Heidegger distinguishes between the autonomous artwork as a means for itself and the equipment (Zeug) as a means for certain purposes. Further, investigating the essence of art does not comprise critically appraising artworks, but rather concerns an exploration of the interdependency between artists and their work: *"The artist is the origin of the work. The work is the origin of the artist"* (2002, 1). Heidegger uses this circular expression to both challenge logical reasoning and to create a virtuous circle between artist, artwork, and the essence of art. He outlines this circle hermeneutically by discussing material, transcendental, and liminal features of actual artworks, looking first at their thingly character.

> *"If we regard works in their pristine reality and do not deceive ourselves, the following becomes evident: works are as naturally present as things. The picture hangs on the wall like a hunting weapon or a hat. A painting – for example, van Gogh's portrayal of a pair of peasant shoes – travels from one exhibition to another. Works are shipped like coal from the Ruhr or logs from the Black Forest. During the war Hölderlin's hymns were packed in the soldier's knapsack along with cleaning equipment. Beethoven's quartets lie in the publisher's storeroom like potatoes in a cellar. Every work has this thingly character. What would they be without it?"*
>
> (2002, 2, 3)

The term "thing" denotes here not just a mere "object" as, e.g., in the natural sciences; a thing in Heidegger's sense is a bearer of different characteristics, a unity in plurality, and formed material.

The difference between natural things (Gk. *physis*) and equipment (*Zeug*), (Gk. *techne*), linking the latter to artworks and design, is explained by referring to van Gogh's *Pair of Shoes* (van Gogh Museum, Amsterdam, 2024), painted in 1886. The picture is presented in *The Origin of the Work of Art* as an example of an artwork as equipment. It illustrates a pair of shoes; however, it transcends this material illustration by evoking images of a lifeworld where these shoes might have been used. It narrates, among others, a farmer's life, working conditions, worries, wishes, and efforts. As an artwork, it allegorises the usefulness of the equipment (the shoes), simultaneously communicating aspects such as its reliability, drudgery, hope, etc., symbolically, thereby transcending material aspects. The symbolisation of reliability, etc., implies first that the creator and the recipient of the artwork are both embedded in a lifeworld where the equipment is used, and second, they have access to an earth, i.e., the material conditions (*nature-physis*), which belong to this world and from which the equipment is made. The equipment in the picture (the shoes) is, however, not revealed through production or utilisation but through artistic illustration, disclosing the being of the artwork:

> *"The equipmental being of equipment was discovered. But how? Not through the description and explanation of a pair of shoes actually present. Not through a report on the process of shoemaking. And not through the observation of the actual use of shoes as it occurs here and there. Rather, the equipmental being of equipment was only discovered by bringing ourselves before the van Gogh painting. It is this that spoke. In proximity to the work we were suddenly somewhere other than we are usually accustomed to be".*
>
> (2002, 15)

The polyvalent potential of an artwork lies, in my opinion, not only in its capacity to evoke diverse material, instrumental, socio-cultural, and ethical dimensions across time and space, but also in its ability to provoke a breakdown, a rupture in habitual perception or interpretation.

A compelling example is René Magritte's *The Treachery of Images* (1929), which famously subverts representational logic through its paradoxical caption: *"Ceci n'est pas une pipe"* ("This is not a pipe"). The picture confronts the viewer with the instability of the relationship between signifier and signified, exposing an ontological gap between representation (the image) and reality (the object it depicts). Similarly, Smith's *This Is Not a Book* (2009) invites readers to reflect on the ontology of the book itself its "book-ness" thereby destabilising conventional notions of textuality, authorship, and the act of reading.

The hermeneutics of Dasein, i.e., human understanding and reflections of beings, as described by Heidegger, comprises reflecting on equipment, while the equipment "talks back"[32]. Within this "conversation", truth arises via the dialectics of the disclosing and the concealing of the object. These dialectics express themselves as "world" (as disclosing) and "earth" (as concealing) and concern all encounters with the lifeworld. An artwork discloses, for example, "world" via connotations and context references, simultaneously creating "earth" via using materials necessary to accomplish the work. Earth is here interpreted generically as protection, and even if it supplies all beings with material that keeps them alive, it is neither material nor the beings themselves, but the unknown essence and source of every being, nature's mystery. "World" denotes actual and potential relationships between humans and other beings, and humans interpret themselves and others within these relationships, thereby expanding their horizon of experiencing and understanding existence (e.g., via equipment breakdown and subsequent change of relationships).

Earth and world are inseparable and cannot exist without one another, while interactions between them remain dynamic, contingent, and context dependent. Heidegger explains the dialectics of earth and world, between disclosure and revealing, with help of artworks. The artwork allows the truth of beings to set themselves to work, fathomable for both the viewer and the creator when contemplating how these dialectics emerge. A non-representational artwork, a Greek temple, illustrates how truth reveals itself in a lifeworldly setting. The temple is important for humans, and this importance manifests itself in a building as cultural, moral, aesthetic, symbolic, and spiritual room. A significant meaning of the temple was to keep the human world open to the gods' presence through a material artwork to a certain group of humans in a certain cultural period. The importance of the temple as a living work of art disappeared, however, when faith in the gods faded, and thus the temple was reduced to an architectural work of art. Heidegger comes back to truth revelation as emerging dialectics of earth and world in his text *Building Dwelling Thinking*, describing a farmhouse in the Black Forest:

> *"Let us think for a while of a farmhouse in the Black Forest, which was built some two hundred years ago by the dwelling of peasants. Here the self-sufficiency of the power to let earth and heaven, divinities and mortals enter in simple oneness into things, ordered the house. It placed the farm on the wind-sheltered mountain slope looking south, among the meadows close to the spring. It gave it the wide overhanging shingle roof whose proper slope bears up under the burden of snow, and which, reaching deep down, shields the chambers against the storms of the long winter nights. It did not forget the altar corner behind the community table; it made room in its chamber for the hallowed places of childbed and the 'tree of the dead' – for that is what they call a coffin there: the Totenbaum – and in this way it designed for the different generations under one roof the character of their journey through time. A craft which, itself sprung from dwelling, still uses its tools and frames as things, built the farmhouse".*
>
> (1971, Section 2)

For Heidegger, art is history, and as historical event, artworks creatively conserve truth. Truth is hence not an absolute, abstract, or metaphysical concept but unfolds itself in time-and-place-specific cultural and historical settings.

Heidegger's philosophy holds relevance for sustainable design in at least three key dimensions. First, it encourages a recognition of nature's *onto-material* value which reflects an intrinsic value that goes beyond anthropocentric notions of utility. This value aligns with a heuristic of *intrinsicality* [33], affirming the value of nature in and of itself. Second, it supports crossing boundaries between material and immaterial aspects of nature, reminding of the Aristotelian concept of *peri ti*, the "in-between" (Keitsch 2003a, 40). The transitional movement invites encounters with nature that are unfamiliar, open-ended, and potentially transformative. Third, Heidegger's hermeneutic approach foregrounds the cultural, temporal, and spatial contexts that continually shape how meaning is ascribed to both objects and the natural world.

In this sense, sustainable design becomes a practice of interpretive engagement rather than mere functional intervention. Heidegger's philosophical insights, particularly his notions of *being-in-the-world*, the *lifeworld* (*Lebenswelt*), and the dynamic interrelations among beings, have significantly influenced various approaches to sustainable and human-centred design. His emphasis on dwelling, authenticity, and place has informed Biophilic Design concepts, which seek to enhance human well-being through reconnection with nature and the built environment. These contributions underscore Heidegger's importance in recognising the experiential and existential dimensions of human–nature relationships.

However, Heidegger's work has also drawn negative appraisal from posthumanist authors, such as Iveson (2015), who criticises its metaphysical anthropocentrism namely, Heidegger's disclosure of Being through a purely human lens. Leach (2016) argues that Heidegger's framework insufficiently accounts for nonhuman agency, thereby reinforcing a subject–object dualism.

Reinterpreting, merging, and deconstructing phenomenological and postmodernist ideas, new materialism (NM) emerged in the 1990s with the aim of developing an object-oriented ontology by removing ontological distinctions between beings. Proponents advocate NM as *material ecocriticism*, while opponents label it *anthropomorphic ventriloquism*, accusing it of projecting human-like agency onto objects. The following chapter discusses the main NM concepts and pinpoints both possibilities and challenges for environmental ethics and sustainable design.

3.5 New Materialism

Philosophically, posthumanism and new materialism (NM) are part of the "Speculative Turn" which emphasises the autonomy of reality from human thought, claiming that the world exists independently of our perception or conceptual frameworks. (Bryant et al., 2011). NM aligns with posthumanist concepts in rejecting the humanist project *per se*. Both theories contest, among other things, that humanist ideas claim to be universal, but this universality refers to a European and Anglo-American, white, male, and rationalist protagonist, which appears absurd in the age of global migration, diversity, and multiculturality. Some NM authors view constructivism and language as a reductionist straightjacket bending the material world for human needs and purposes; others attempt to merge language and materiality in a "material discursivity". NM also discards anthropocentric perspectives of the nature–culture divide and the idea of the subject as *"all-knowing, all-seeing agent of history"* (Soper, 2012, 369). Another common ground for posthumanism and NM is the proclamation that the ontological difference between beings is obsolete, and that "entanglements" between humans and non-human entities, e.g., technologies, objects, and environments, contribute to the construction of subjectivity and social structures. One of the first works to address this disintegrated ontology is Haraway's *A Cyborg Manifesto* (2016a), presenting the cyborg as a metaphor for the blurred boundaries between humans, machines, and nature. Hardt et al. (2000) explain further:

> *"human nature is in no way separate from nature as a whole, that there are no fixed and necessary boundaries between the human and the animal, the human and the machine, the male and the female, and so forth; it is the recognition that nature itself is an artificial terrain open to ever new mutations, mixtures, and hybridisations".*
>
> (2000, 215)

Barad's concept of posthumanist performativity holds a central position in NM theories. Performativity refuses the idea that words represent "things" (representationalism), and Barad (2003) argues that posthumanist performativity must include material, discursive, social, scientific, human, nonhuman, natural, and cultural aspects, but more importantly, she insists that boundaries between beings are per se blurred and that interdependency and bonds between beings should be acknowledged as a starting point for theories. Barad's "agential realism" emphasises the entanglement of processes and matter, instead of focusing on construction and interpretation. Agential realism understands phenomena as correlations between objects and their measuring agents (2007, 337) emerging through intra-action: *"the universe is agential intra-activity in its becoming. The primary ontological units are not 'things' but phenomena – dynamic topological reconfigurings/entanglements/ relationalities/(re)articulations" (Barad, 2003 , 818). "A phenomenon is a specific intra-action of an 'object' and the 'measuring agencies'; the object and the measuring agencies emerge from, rather than precede, the intra-action that produces them"* (Barad, 2007, 128).

The dynamic and processual nature of intra-acting agencies by constituting each other in an intra-action thereby delimits the borders to each other. Within this interpretation, every research object is, for example, entangled in the way of investigation, which opposes traditional science with its predetermined attributes and boundaries of objects. The aim of new materialism is to reconceptualise relationships between humans and non-humans and to revise categories of subjectivity, action, and causality. The material-discursive approach explains, for example, that materials carry the historicity of practices in themselves (Barad, 2007). Barad is not talking about dendrochronology when she refers to, e.g., growth rings as embodied history of a tree (2007), but the rings illustrate embodied ongoing intra-active relations.

Barad and other NM concepts stress the materiality of reality; however, this materiality seems inevitably aligned to human experiences, perceptions, and practice. While the idea that the observing, interpreting subject (temporarily, M.K.) dissolves in intra-actions allows new liminal ways of experiencing phenomena, it also comes with ethical challenges, e.g., when neglecting the dynamic relatedness of human experiences and decision-making.

The ethical approach in NM searches for less or non-anthropocentric concepts, for example by emphasising compassion for the common vulnerability and thrownness[34] that all beings share. From an environmental ethics perspective, this compassion and empathy are important stimuli for ethical reflection and potential action. Yet, some authors find the anti-dualist position, represented among others by a lack of distinction between humans and nature, problematic (Rekret, 2016; Žižek, 2014; Herbrechter, 2013; Soper, 2012). NM is, among others, criticised for its tendency toward "ecological naturalism" (Soper, 2012), which risks collapsing the distinction between humans and nature by emphasising human dependence on other life forms and ecological systems. This position encounters similar aporias to those explored in Chapter 2 of this volume, particularly in relation to "Biocentrism" and "Deep Ecology".

Regarding materialism, Žižek (2014) points, for example, to the ambiguity of assigning general value to materials because of their vital qualities: *"are these … the result of our (the human observer's) 'benign anthropomorphism', so that the vitality of matter means that 'everything is, in a sense, alive', (quote Bennett, 2007, 117) or are we effectively dealing with a strong ontological claim asserting a kind of spiritualism without gods, with a way of restoring sacredness to worldliness?"* (Žižek, 2014, 13). Earlier, he appraised NM as pre-critical, naïve realism: *"We … move back to the enchanted world"* (Žižek (2014), 12).

Other authors (Rekret, 2016; Soper, 2012) mention that NM's environmental egalitarianism, ascribing equal value to all existence, diminishes basic ethical ideas such as "responsibility". In their view, the human condition comprises moral sensibility and responsibility *sui generis*, and these include concern and care for other beings and ecosystems as well. In NM, human responsibility is suspended in a fusion of ontology, epistemology, and ethics. For Barad, "responsibility" is, for instance, inherent in any relationship, and rights and obligations do not apply only to humans, but to all beings.

The absence of ethical and political differentiation in NM has been identified as a significant limitation, particularly in relation to emancipatory and critical discourses (Soper, 2012). Without clear distinctions between ethical values in relational configurations, it becomes difficult to evaluate which intra-actions are more desirable, just, or beneficial than others. Moreover, existing exclusion and power asymmetries within societal groups, as well as between humans and non-humans, such as how material is distributed and accessed and by whom, are hardly thematised in NM (Soper, 2012).

The significance of NM for sustainable design lies in its potential to adopt a perspective of nature as vital and processual, rather than merely objective, predetermined, constructed, or solely semantic. Designerly interpretations of a "multispecies or inter-species methodology" are explored in more depth in subchapter 5.2, with some examples provided below. Contemporary design concepts that incorporate NM ideas often emphasise the intrinsic connection between knowledge and practice, which are fundamentally intertwined, as Barad, for example, remarks:

> *"Practices of knowing and being are not isolable; they are mutually implicated. We don't obtain knowledge by standing outside the world; we know because we are of the world. We are part of the world in its differential becoming. The separation of epistemology from ontology is a reverberation of a metaphysics that assumes an inherent difference between human and nonhuman, subject and object, mind and body, matter and discourse".*
>
> (Barad, 2007, 185)

The idea of connecting epistemology and ontology is not novel; for example Kuhn (1962), argues that scientific thinking (epistemology) is always part of a scientific community, which has to respond to societal forces such as polity, economy, and culture (ontology). Even if the relationship between ontology and epistemology is a contested issue in philosophy of science, many authors refuse to conflate epistemology and ontology. First, because a fusion logically evokes "category mistakes"[35]; second, because a researcher's semantic understanding of ontology and epistemology, acknowledged or not, necessarily influences the way theory and practice are approached (Marsh and Stoker, 2010).

Despite the criticism that NM rejects standard knowledge generation but is vague in explaining alternatives, NM concepts stipulate a fruitful material–practice relationship for design research and practice by *"turning matter into medium"* (Büsse, 2020, 355). Büsse advocates, e.g., turning to a *"more-than-human-centred design"* (2020, 359) by including materials. Implications of the latter for design relate materials then to ethical questions such as *"who can design, to the benefit of whom, and towards which ends?"* (2020, 360). Büsse emphasises the impact of designed materials, because discussing impacts opens new reflections and discourses in design theory. Material loses its passive and instrumental role and advances to a medium for intra-actions that unites societal and material factors alike, illustrated with a project on sand's entanglements and transformations in Indonesia, Malaysia, and Singapore. The project shows connections between humans and the environment, exploring material beyond its mere physical attributes. Büsse asserts that sand has a plastic capacity as forming and being formed; it is *"neither inert, awaiting to be formed by design, nor is it active and flowing without restraints"* (Büsse, 2020, 369). Either way, taking form and giving form, Büsse understands material not as formless, but as permitting a "plastic reading" as a "material-discursive" (Ulmer, 2015) reading. In ecology, the term "plasticity" denotes the behavioural change of an organism adapting to altering environmental conditions within its lifespan (West-Eberhard, 2003). NM emphasises the mutational and transformational attributes of plasticity metaphorically as ontological ways of being and becoming, e.g., in Malabou's (2022) concept of plasticity.

The perceived "plasticity of materials" facilitates sustainable design by widening perceptions of natural and designed objects, for example by including their embodied history.

Another recent and designerly valuable component of NM sees relationships with materials through the lens of material entanglements (Jones, 2024). Jones investigates laypersons and knitting as an everyday activity. Some of those reflections also apply to how designers engage with materials professionally, such as the remark that the significance of an object is not the sole result of cognitive and linguistic processes, even if those are favoured in disciplinary curricula, but also concerns emotional, aesthetic, and semantic bonds. Referring to Barad's agential realism: *"Meaning is not a property of individual words or groups of words but an ongoing performance of the world in its differential dance of intelligibility and unintelligibility"* (Barad, 2007, 149). Jones' basic idea is, in my opinion, that materials inhere change and development in the form of various entanglements (instrumental, aesthetic, social, etc.), which contribute to different degrees and within varying situational settings to engagement. Within the *"literacy in everyday life"* (Jones, 2024, 83), the knitting process is described dynamically with special focus on "becoming", i.e. the transformation of the material through designerly intervention: *"how knitting comes to matter for makers"*

(Jones, 2024, 84). The knitting material is not separated from activity, process, and elements (strings, stitches, rows) but presented as an intertwined, interdependent pattern, a correspondence between designer and designed, all of which contribute to "meaning making". Jones presents helpful insights on the composite nature of knitting, which could be similarly explored for drawing, sketching, and other professional design activities. Knitting elucidates how participants in the case study valued, for example, that *"the materiality of emerging knitted fabric offered a visual representation of time and effort"* (2024, 10), and, regarding the latter, how produced objects represent an appreciation of relationships, e.g. as a gift. This provides important moments of emerging novel relationships with the material, and with oneself:

> *"I want it (knitting, M.K.) to be a pleasure, otherwise I just won't do it. I don't finish things. And also, I happily leave things. I've certainly finished things after six years, and I don't think it matters. It doesn't. That's fine".* (2024, 15)

Embracing imperfection, regarding the human–nature relationship, can contribute, among other things, to reversing reductionist and idealistic perceptions of nature as a flawless maker, compelling humans to achieve the same flawlessness in design. Practically, human–nature connections and material-discursive entanglements are included in, e.g., the "Connectedness with Nature Concept" (Veslova, 2019). Veslova (2019) recommends considering sustainability interactions in human–nature systems in every project, recognising all actors (also natural entities) as stakeholders. Conclusively, new materialism-inspired approaches, such as multispecies ethnography and material–discursive entanglements, can enhance sustainable design. As discussed in Chapter 5.4, they broaden analytical epistemologies by incorporating complex interrelations between humans and nature, offering a valuable supplement to responsive design practices.

3.6 Persuasive Design for Sustainability

The first decades of the 21st century have seen the emergence of numerous sustainable design approaches grounded in diverse philosophical traditions ranging from postmodernist, speculative, and constructivist thought to post-humanism, and more recently, liminality. The latter, a post-phenomenological philosophy, emphasises the contingent, transitional, and in-between nature of experiences and encounters with the natural world, focusing on *natura naturans* rather than *natura naturata*, so to speak.

Chapter 4 explores a distinct yet influential approach persuasive design and technology which draws on a behaviourist paradigm and empirical psychology to promote various goals, including the advancement of sustainability.

Persuasive design (PD) attempts to influence users' and stakeholders' actions, for instance by nudging environmental behaviour. It is often argued that PD strategies can support more sustainable consumption of products and services by fostering both positive emotional responses and heightened environmental awareness. However, persuasive design (PD) has also raised concerns about its ethical implications, requiring thorough, context-sensitive evaluations of both the design and the intentions behind persuasive solutions.

The most noteworthy philosophical foundation of persuasive design is Behaviourism, which observes and analyses stimuli that affect organisms, animals, and humans, and the reactions these stimuli trigger. Epistemically, behaviourism considers primarily the empirical evidence of an individual's behaviour. Mental processes or emotions are internal, neither observable nor testable; they do not contribute to generating knowledge and are hence regarded as irrelevant, "empty" (Carnap, 1931). Empiricists argue that knowledge arises from observation and experimentation, which are used to validate or falsify hypotheses unlike classical rationalism, which relies on a priori reasoning. From an empiricist standpoint, the idea that understanding involves not only describing reality but also pursuing non-empirical or transcendental goals – such as the purpose of research or the meaning of knowing – is regarded as metaphysical speculation, and thus, meaningless within an empirical framework. Psychological behaviourism adopts the study of sensory experiences of animals and humans empirically as a basis for knowledge generation, assuming that sensory experiences construct personality, character, beliefs, emotions, and behaviour. Desirable behaviours are either achieved through coupling stimulation and behaviours or by rewarding behaviours. One of the first behaviourists, Pavlov (1928), studied, for example, the so-called "conditioned reflex" of dogs, an unconscious reaction to a stimulus, and investigated how a different stimulus can produce the same reaction if the study object is conditioned to associate it with the original stimulus.

> *"The procedure of development of such reflexes is as follows. Any convenient external stimulus is applied to the animal and continued for to 1 minute. After another definite interval of 1–3 minutes, food or a rejectable substance is introduced into the mouth. It is found that after several repetitions of this routine the stimulus will not itself evoke any reaction; neither will its disappearance; but the appropriate reaction will occur after a definite interval, the aftereffect of the excitation caused by the stimulus being the operative factor".*
>
> (1928, 39)

Society plays a crucial role in early behaviourism concepts as well. Mead (1912) claims, e.g., that an individual surfaces only in social relations and through intersubjective interactions, internalising ideas, opinions, and feelings connected with these. Skinner (1973) examines the conditioning of human behaviours in the form of social rewards or rejection. A stimulus elicits a reaction, which can then be positively or negatively reinforced: *"Negative reinforcers are called aversive in the sense that they are the things organisms 'turn away from'*" (1973, 32). Positive reinforcement relates to experiencing a desirable or pleasant stimulus materially, physically, socially, or combined:

> *"We applaud a performer precisely to induce him to repeat his performance, as the expressions 'Again!' 'Encore!' and 'Bis!' indicate. We attest to the value of a person's behaviour by patting him on the back or saying 'Good!' or 'Right!' or giving him a 'token of our esteem' such as a prize, honour, or award".*
>
> (1973, 48)

An important psychological element for persuasive design is the "Theory of Planned Behaviour", claiming that individuals have intentions which lead to behaviour, but that realising this behaviour depends on the ambition to do so. According to Ajzen et al. (2009), attitudes towards a behaviour, such as expectations of a positive outcome, subjective norms such as peer pressure and normative beliefs, and perceived behavioural control regarding resources, obstacles, and power issues influence the intention. Employing this slightly rationalist picture of the mental abilities of a *homo economicus*, Fogg (2009) aims to automate behaviour with the help of persuasive technology to mitigate the intention–behaviour gap. His behavioural model (FBM) comprises three factors: motivation, ability, and trigger, that must converge concurrently to realise behaviour. The right dosages and types of interference are crucial to success. Fogg describes interferences for all factors in detail; for motivation, these relate to pleasure and pain, hope and fear, and social approval and rejection. Enhancing ability, elements of simplicity are crucial.

> *"In real-world design, increasing ability is not about teaching people to do new things or training them for improvement. People are generally resistant to teaching and training because it requires effort. This clashes with the natural wiring of human adults: We are fundamentally lazy".*
>
> (2009, 6)

Via simplicity, persuasive design can improve abilities to achieve a behaviour by, e.g., reducing time, costs, and physical or mental effort. Finally, triggers prompt the performance of the behaviour. Hot triggers allow immediate action (e.g., a pop-up button "save energy now"); cold triggers cannot be acted upon instantly (e.g., a brochure on how to save energy in a household).

Persuasive design, aligned with user groups' values, abilities, and motivations, has also become popular in supporting lifestyle choices, e.g. adopting and maintaining positively perceived habits or desirable behaviours. Gamification elements, inbuilt feedback from a "community", incentives, and rewards are all design elements to induce behaviour. Disadvantages of persuasive design include manipulation, lack of user awareness of the persuasive technology in applications, and discrepancies between profit orientation and values of social and individual well-being (e.g., doomscrolling). Reduction of critical thinking, overreliance on guidance instead of personal decision-making, and blurred boundaries between coercion and nudging are other risks.

Before addressing the ethical challenges, the potential of persuasive design to support sustainability is illustrated through the concept of "nudging". Nudging seeks to influence behaviour without pressuring or restricting an individual to change but by changing how choices are presented. Sunstein (2015) explains nudging basically as influencing decision-making. His all-encompassing view of nudging even includes natural phenomena to justify why nudging is inevitable: "Nature nudges" (2015, 421). Yet, if nature nudges, some sort of intention or purposefulness must be noticeable, which is difficult for several reasons discussed in earlier chapters of this issue. Due to human unfamiliarity with "nature's purposes", the societal interpretation of nudging as an inherent constituent of interactions is more plausible than Sunstein's naturalist position. Schmidt et al. (2020) connect nudging, e.g., with "choice architecture", and Talgiabue and Sandaker (2019) emphasise that choice architecture is embedded in natural, social, and historical (M.K.) settings, which influence behaviour. Most authors confirm that deliberate influence on how choices and options are framed and presented to individuals can potentially transform behaviour. Some authors, however, overgeneralise nudging, claiming that it cannot be avoided anyway because it is inherent in all social interactions (see also the naturalist justification). Both credos that nudging is immanent in every social interaction, and that decisions made instinctively rather than deliberately, is epitomised by Thaler et al.: *"First, never underestimate the power of inertia. Second, that power can be harnessed"* (Thaler et al., 2008, 8).

Yet choice architects, who design interventions, seem to engage in deliberate, rational decision-making rather than relying on instinctive or habitual judgments (see also "social engineering" in subchapter 2.5). This exposes a profound asymmetry: while end-users are steered through cues that intentionally capitalise on cognitive shortcuts and automatic processes, the designers of these choice environments employ calculated, reflective reasoning to forecast outcomes which in turn raises important concerns about designers' agency and decision-making power. It also underscores the inherently normative dimension of their role, positioning them as active agents in constructing contexts and thereby subjecting their decisions to heightened ethical and political scrutiny.

Nudging is employed in both the public and private sectors, and governmental nudging is seen as valuable for policymaking in areas such as the public sphere, healthcare, and the environment (Schmidt and Engelen, 2020). In domains where decision-making has important societal consequences, nudging can help to realise ends that have been chosen (for example via public decision-making, M.K.) by improving the means to achieve them.

> *"This argument becomes stronger if you consider how many of our environments are shaped by private companies that use choice architecture ill-matched to our decision-making procedures. Food companies, for example, systematically trigger and exploit our psychological dispositions so that we*

> *overeat and overspend. Correcting such environments through public policy nudging, or so the argument goes, could counter this, and achieve a better match between our decision-making propensities and our choice environments".*
>
> (Schmidt and Engelen, 2020, 6)

Several authors agree that increased global political crises, environmental hazards, and the development of innovative technologies all significantly change societal and individual values, and that, consequently, nudging strategies become more relevant in political-normative programmes, e.g. in transforming sustainability practices (Knowles et al., 2016; Talgiabue and Sandaker (2019; Schmidt et al., 2020).

Nudging for sustainability is based on anthropocentric premises; nature is primarily assigned instrumental value as a resource and as a means to support human well-being. On a micro- and meso level, various biophilic nudges are designed to infuse experiences of being close to nature. Natural light, immersive backgrounds that change with the seasons; calming nature sounds; natural materials and geometries; and natural phenomena as navigation metaphors (rivers, paths, clouds) are all designed to evoke psychological reactions and positive attitudes towards nature and, indirectly, sustainability. On a meso- and macro level, governments implement nudging as a public policy instrument to promote pro-environmental behaviour, mainly via information dissemination, changes in the physical environment, changes in setting default options (i.e. actions conducted if nothing else is specified by the user), and consumption habits feedback (Nielsen et al., 2016).

According to Sunstein (2016), libertarian nudges allow citizens to decide freely, while paternalistic nudges prod them into making better decisions. Authors also recommend that systemic nudges should address at least two thinking levels: (1) fast: automatic and intuitive; and (2) slow: deliberate and conscious (Kahnemann, 2011). Effective nudging interventions for sustainability on level 1 can be implemented through, e.g., defaults such as a preset double-sided print option in offices, and these nudges should be visually noticeable, attractive, and easy to perform. On level 2, governments can address attitude-behaviour gaps, i.e. expressing positive attitudes towards sustainable actions without acting accordingly, via information, campaigns, eco-labelling, etc. (Sunstein, 2016). An advantage of government nudging is its potential to guide citizens in decision-making processes, providing possibilities for informed choices. Congiu et al. (2021) distinguish between "pro-self nudges", facilitating individuals to improve behaviour for their own long-term well-being, and "pro-social nudges" that aim at the welfare of society by reducing behaviours that diminish common well-being.

Nudging literature largely agrees that interfering with citizens' sustainable behaviour must be carried out by democratically elected representatives, and that decision-makers and designers of nudges, persuasive technology, and choice architecture must follow ethical principles. Many authors discuss ethical challenges connected to nudging and develop strategies to mitigate these.

The fact that nudging is social engineering, and nudges are purposely designed alterations of behaviour, remains, however, the bottom line, even for well-meant utilisations. Even in cases of self-tracking to voluntarily change towards sustainable behaviours, actions are delegated to a device, leaving choices to sensors and algorithms, which raises, among other things, the question: At what point does the autonomous decision become automatised? (Doyuran, 2023).

Another substantial challenge for nudging is that users can be manipulated without their knowledge and informed consent. This relates to a potential lack of transparency in disclosing design mechanisms that are influencing behaviour. Privacy concerns about collecting, storing, and processing user data are another ethical issue for nudging, persuasive technology, and choice architecture. Some authors also criticise a lack of cultural sensitivity; Schmidt et al. (2020), for example, recommend evaluating nudges within their specific contextual settings, while Kimura and Nakajima. (2011) explore ways to design nudges that are culturally appropriate. Clavien (2018) develops a framework to evaluate whether nudges are ethically appropriate, with the following guidelines for a designer or nudger: to justify motives and good intentions; to demonstrate trustworthiness, i.e. provide validation that there is no conflict of interest; to empower democratic institutions and include relevant stakeholders in the nudging design; and finally, to ensure that the nudge has no or few unwanted side effects. Additionally, nudging, and persuasive design and technology should actively facilitate societal context-specific values which express concern for the community and the environment and foster political engagement (Knowles et al., 2016).

The challenges above indicate that much of the responsibility to align persuasive design, nudging, and choice architecture with ethical principles is transferred to designers. The "Golden Rule" of persuasive technology is that: *"Designers ought not to persuade users to do something that they themselves would not wish to be persuaded to do"* (Knowles et al., 2016, 60). This categorical imperative assumes, first, that designers know what is good for them, second, that they act accordingly, and third, that they can predict what is good for the users. Even if all of this is hypothetically the case, differences between designers' subjective ethical values and professional responsibilities can be huge, and ethical dilemmas arising from these differences are both structural (relating to the professional codex) and personal (relating to one's own lifestyle, culture, etc.). The need to socially belong to a group by displaying "correct" behaviour can, for instance, be misused to nudge consumers into believing that representatives of this group also purchase and like a product (following the professional credo of "selling products"). As professionals working in industries that aim to maximise profits, designers are part of a structure that promotes excessive spending rather than sensible consumption, and persuasive design tools and strategies, not immune to misuse, allow for manifold types of exploitation.

Persuasive technology, design, and nudging matter for political decision-making as well. Governments are recommended by authors to stand last in a line when stakeholders and/or the public have discussed and revised, for instance, environmental targets to evaluate whether they are still sufficient for contemporary challenges and have decided upon socially appropriate solutions. Applying the principle of proportionality, governments must then select means for implementation, choosing the least restrictive means regarding individual conditions (Treaty on European Union, 2023). If persuasive design is implemented post hoc in contextualised settings, it can facilitate strategies to approach selected solutions. Implementing nudging on a widespread public scale – "Big Nudging", i.e. the combination of nudging with big data (Helbing, 2017) is, on the other hand, seen as both problematic for citizens' autonomy and politically counterproductive in addressing complex societal problems. According to Helbing (2017), employing identical measures across an entire population can consequently prove more harmful than beneficial for society:

> *"In fact, 'Big Nudging' aims to synchronize individual actions and the manipulation of perspectives and decisions. This comes close to a targeted incapacitation of citizens through state-planned behaviour control. We fear that the effects could be fatal in the long term".*
>
> (Helbing et al., 2017, 18, trans. M.K.)[36]

.

In summary, nudging can promote sustainable consumption and reduce ecological impact through behaviourally informed interventions in sustainable design (Tagliabue et al., 2019). However, these strategies raise ethical concerns when they compromise user autonomy by operating without transparency or informed consent (Schmidt et al., 2020).

Autonomy plays a central role for both humans and nature in ethical discourses. The notion of autonomy is revisited and examined in greater detail in the following chapter, which explores its relation to aesthetic experience and whether this relation can foster environmental responsibility and inform sustainable design practices.

3.7 Aesthetic Experiences, Environmental Ethics, and Sustainable Design

The idea that aesthetic experiences are fundamental for ethical actions reaches back to ancient philosophy, and it might be relevant to take a small detour to explore some origins of contemporary concepts. Plato's and Aristotle's ideas on beauty as a well-structured whole, harmony, practicality, regularity, and the alignment of opposites are essential for early aesthetics-ethics relations. The leitmotif in Plato's *Philebus* (4th century BCE) is that the beautiful and the good are identical: *"now we notice that the force of the good has taken up refuge in an alliance with the nature of the beautiful. For measure and proportion manifest themselves in all areas of beauty and virtue"* (Plato, 2008, 64d–65a,). For Aristotle, the trinity of the true, good, and beautiful is present in all nature, artefacts, and morally appropriate actions. In the 3rd century CE, Neoplatonism with Plotinus becomes the harbinger of modern subjectivism and aesthetic experience (of nature). For Plotinus, the individual aesthetic experience of sensual beauty rises into a self-reflective, anamnestic (remembering) recognition of the beauty of the soul and the spirit (God). Inner contemplation must be stimulated morally through asceticism, meditation, and contemplation, and if successful, culminates in an experienced union with God, the *unio mystica*. Plotinus' philosophy advances that beauty is not something sinful in a "valley of tears" called world, but a manifestation of God and of nature, his creation, which sanctions beauty morally because it expresses God's creation through unity, proportion, and harmony. The Renaissance interprets aesthetic experiences (of nature) as a philosophical endeavour to cultivate lyrical sensitivity, feelings, and inner contemplation. Petrarch (1304–1374), with a description of his ascent to Mont Ventoux, stands at the beginning of European literature, narrating landscape experiences. Highlighting the positive value of aesthetic experiences for emotional life, Petrarch detaches landscape from instrumental usability and attaches it to mental development (Keitsch, 2003a).

Nature as a metaphor for human autonomy embodies the Enlightenment ideal of aesthetic experience, linking freedom, reason, and sensibility in a unified vision. Cultivating one's own humanity via aesthetic experiences culminates in the 17th century, among others, in emotivism, such as Shaftesbury's theory of a beautiful life, lived with moral grace, where aesthetic experience and moral sense correlate with sociability and enthusiasm for living nature, all flowing into an ethics of feeling (Gill, 2022).

A philosophical summit of reflecting on the relation between aesthetic experience and ethical conduct marks Kant's *Critique of Judgement* (1790), a metaphysics of natural beauty and the sublime, written as a supplement to his *Metaphysics of Morals.* Kant applies the term "aesthetics", coined and explained by Baumgarten in 1750, to the entire field of sensory experience, and rejecting Utilitarianism and Emotivism, he finds morality and aesthetics in reason. Anthropologically, rationality and autonomy distinguish humans from other living beings; morally, they oblige humans to do justice to their own nature and act as rational beings. The *Critique of Judgement* explains rationality and autonomy via aesthetic experiences.

First, experiencing the phenomenal sublime in external nature, the vastness of the sea, the power of the mountains, represents for Kant an equivalent of the inner experience of freedom. The beginning of the conclusion in *Critique of Practical Reason* (1788) reads: *"Two things fill the mind with ever-increasing wonder and awe, the more often and the more intensely the mind of thought is drawn to them: the starry heavens above me and the moral law within me".* Experiencing nature's sublimity requires a distancing from one's physical existence and immediate needs, thereby opening a space for the recognition of individual freedom.

> *"Hence ... we call it sublime, we do so not because nature arouses fear, but because it calls forth our strength ... it elevates our imagination, it exhibits those cases where the mind can come to feel its own sublimity, which lies in its vocation and elevates it even above nature."*
>
> (*Critique of Judgement* §28)

Second, natural beauty inspires humans and awakens moral feelings because moral judgments and aesthetic judgments have the same formal condition, namely freedom. Natural beauty provides an ideal context for the cultivation of moral capacities, as it engages the mind in ways that encourage moral reflection. Aesthetic experience and aesthetic judgment work autonomously, remaining independent of external stimuli or personal desires. Beautiful creatures, artworks, and objects do not only evoke joy, but they can stimulate the experience of freedom, for example as detachment from mundane daily goal settings. Aesthetic judgment sees the world from its own perspective and shows neither practical interest in the perceived object nor theoretical interest in its concept. The beautiful hence differs from the pleasant because it evokes "uninterested appreciation" thereby resembling "the good" which is also done for its own purpose in Kant's deontological ethics. As a moral individual acts according to the categorical imperative without paying attention to personal interests and thereby experiences freedom regardless of sensual and bodily determination, an aesthetic observer looks at the beautiful without intent. Aesthetic objects remain inexplicable: the beautiful encourages thinking without being itself subject to the criteria of theoretical and practical reason, and for Kant it therefore resides in an independent dimension besides the good and the true.

Third, aesthetic experience does not relate to object knowledge, but to meaning – *telos.* Kant's aesthetic-hermeneutic (*as-if* [37]) teleology allows one to experience the world as an organic unit in which humans are integrated, evoking, e.g., feelings of harmony, of a world which is functional and beautiful without necessarily serving human purposes.

In the history of ideas, Kant's positioning of aesthetic judgment as a mediating faculty between the realm of nature (guided by theoretical reason) and the realm of morality (guided by practical reason) represents a remarkable philosophical achievement. Emphasising the lack of instrumental interest in beauty, Kant liberates aesthetics from the instrumental grasp enmeshing beauty and advantage, thereby maintaining the non-instrumental value of (natural) beauty. Moreover, the insight that sensitivity to natural beauty and tenderness for other creatures expresses humanity emphasises human aesthetic freedom and morality in a way that gives new impetus to human self-understanding.

Kant's ideas on nature, aesthetic experience, and ethical conduct were adapted and transformed over the following centuries by numerous authors, some of whom have already been discussed in this issue. Environmental ethics and aesthetic nature experiences share provenance and relationships in the history of moral philosophy and environmental aesthetics, a common descendant that emerged in the 1960s. Environmental aesthetics examines human dispositions and preferences for certain types of nature, landscapes, species, etc., whether it is possible to formulate a claim for the existence of these types (e.g. as Natural Heritage), whether these claims can be ethically justified and, if so, what norms result from their justification. Newer environmental aesthetics concepts discard the nature–culture antinomy and consider nature as dependent on spatially and temporally different socio-cultural structures, where actors define what nature "means", which values are associated with nature, and which actions toward nature are appropriate. Most environmental aesthetics concepts address nature experiences within a practical-expressive dimension, i.e. individual experiences of nature are linked to intersubjective interpretations and socio-cultural values and norms. Some authors presume that physical and sensory elements (sensuality concepts) or affective elements (emotive concepts) such as feelings and emotions (Parsons, 2008; Wang et al., 2018) contribute to connectedness, identification, and empathy with nature.

Sensuality and emotive concepts both derive ethical motivations from aesthetic nature experience. Emotions are here seen as relevant for sustainable actions alongside social values and norms, and findings from ecology, e.g. in Wang et al. (2018):

> *"an innate connectedness to nature is the state of mind that one can develop when interacting with nature. Whether it is approached through cognitive, experiential, or belief-oriented paths, the emotional attachment between human beings and the environment is predictor of appropriate ecological behavior".*
>
> (2018, 4)

Authors such as Böhme (2010) and Berleant and Carlson (2007) highlight phenomenological analogies between external nature and the human body as an onset for individual and social moral development. Böhme's anthropological aesthetics aims, for example, to facilitate bodily experiences within a socially constructed nature. Bodily experiences allow one to (re)establish tangible relationships with outer nature via the immediate feeling of one's own body within an environment. Sensual sensitivity evokes feelings of belonging and facilitates, eventually, a sense of responsibility toward one's own and the "other" nature. The connection of bodily experiences and emotive reactions culminates in Böhme's concept of nature as a place of feelings (of one's own body and for external nature), which creates an anchor for morality. Böhme's argument of subjective experiences and subsequent emotions and ethical attitudes presumes a contemplative, non-instrumental relationship with aesthetic nature. If contemplative nature experiences are generally evaluated as desirable and beneficial, this implies a moral duty to preserve natural surroundings that allow these experiences as an option for other humans and for a good human life. Aesthetic experiences happen here in an Aristotelian sense;

perceptions of the surrounding nature correspond with human ideas of order, meaning, harmony, proportion, and rhythm, meeting expectations of a meaningful world, where everything has its "intrinsically" right place and is apparently "making sense", including human existence. The correspondence perspective occurs habitually in interpretations of aesthetic nature, for example in Böhme's concept of "atmospheres": *"Atmosphere is the common reality of the perceiver and the perceived. It is the reality of the perceived as the sphere of its presence and the reality of the perceiver, insofar as in sensing the atmosphere s/he is bodily present in a certain way .... A Valley is ... not called serene because it is in some way similar to a cheerful person but because the atmosphere which it radiates is serene and can put this person into a serene mood"* (Böhme, 1993, 124). Böhme states that everyday life already has a variety of characterisations of "atmosphere" sometimes associated with terms such as "ephemeral", "intangible" or "auratic". Existing relationships between natural and cultural settings that manifest atmospheres are, for example, described by the term *genius loci* (spirit of the place)[38].

The atmosphere concept is regularly reflected in contemporary sustainable design, planning, and architecture. Even if an atmosphere itself cannot be designed, *"rather, the role of design is to create interventions that make possible the circumstances through which particular types of atmospheres might emerge, but alone it cannot predetermine or predict exactly what these atmospheres will be experientially"* (Sumartojo et al., 2019, 95), creating atmospheres remains an option to develop space for aesthetic experience. For Böhme (2017), atmosphere creation happens at the interface between acknowledging natural states, anthropomorphic nature interpretations, and understanding cultural needs and dispositions. Illustrating relationships between subjective experiences and phenomena of the external world, atmospheres are not merely subjective projections, but surface through an interplay between individuals, objects, and surroundings, producing, for instance, moods to be in a certain place or avoid it. One goal of creating atmospheres is, for Böhme, to generate feelings and emotional spaces, and in contrast to everyday life where atmospheres often arise randomly, professionals should be aware of the power of atmosphere design and the implied responsibility (although Böhme is not entirely clear whether this responsibility refers to ethical or aesthetic criteria, or both).

Designing desirable atmospheres thus becomes ethically relevant for an expanded discourse about nature, considering generic needs and values for a certain type of nature experience. However, assigning specific atmospheres to explicit moods is also problematic, because it contradicts the indeterminateness, polyvalence, and openness of aesthetic experiences. Further, it is difficult to confirm the exact atmospheric elements that create a mood. On an overarching ethical level, if the assumption that appropriate surroundings stimulate ethical action remains valid, designing atmospheres must support the idea of a good life (eudaimonism).

Practically, the design of atmospheres is, among other things, seen as a supplement to visual approaches that mainly focus on aesthetic proportions, relations among forms, contrasts, shapes, etc., of places (Eronen, 2024), and as an opportunity to facilitate experiences of ecological connections to enhance awareness of ecology (DeKay, 2011). Atmosphere creation targets spaces where material and immaterial aspects are intertwined in multilayered environments (Sumartojo et al., 2019), mediated by objects, human activities, and natural settings. Creating atmospheres for places that represent the aesthetics of daily life and produce emotional relations is seen as particularly rewarding. Place atmospheres are dynamic and change spatially and temporally with stakeholders' activities, settings, and seasons (Eronen, 2024; Singh et al., 2018).

While the hands-on feasibility of the approaches above is clear, discussions of the dialectics between objectively staged and subjectively experienced atmospheres fall short in many profession-oriented concepts. Most focus on augmenting generic well-being in everyday aesthetic experiences and on creating feelings of belonging and being welcome, all supposed to contribute to enhanced life quality (Kuruoğlu et al., 2021).

Still, aesthetic nature experience is not always beautiful, and beautiful nature does not necessarily evoke harmonious and cosy feelings, or social, emotional, or spiritual well-being (Świątek et al., 2023). Ethical reflections and motivations can also be triggered by a different type of aesthetic nature experience, which seems overlooked yet is very relevant for the sustainable design debate on environmental aesthetics.

Nature phenomena that contradict pleasing harmonies, such as in Kant's "sublime" or Adorno's "non-identity", render beauty irrelevant for nature experience. "Disturbing" aesthetic experiences dismiss goal-directedness and purposefulness and challenge arduously cultivated contemplation. The meaninglessness, impermanence, disproportionality, and confusion that these experiences induce convey that nature is "larger than human" (Świątek et al., 2023). According to Welsch (1987), experiencing nature as surpassing human understanding and rules can trigger reflections on nature's independence and sovereignty. Experiencing "disorderly" nature then allows one to face autonomy *ex negativo* as ""*pleasure which is possible only by means of a displeasure*" (Kant, 1790, §27). The insight that nature, although shaped by socio-cultural actions, lies outside human abilities to comprehend, control, and design bears a resemblance to the experience of autonomous art (Keitsch, 2003a)[39]. According to Seel, disturbing nature facilitates experiencing freedom in different ways: freedom from meanings, freedom to explore one's own existence, and freedom to invent new ways of well-being (Seel, 1996). In this way, the "symbolic" autonomy of nature represents a non-human forum for humans to become familiar with themselves and the opportunities to shape their lives (Seel et al., 1992), thereby allowing for individual and collective reflections on a "good life" (in a eudaimonistic sense).

Enabling aesthetic nature experiences is often interpreted as a socio-cultural task (Seel, 1996; Eronen, 2024; Barros et al., 2021; Kuruoğlu et al., 2021). Nature is assumed to be constructed by cultural standards, which in turn would not exist without reference to nature. The ethical implication is that if nature is viewed as an important part of culture, the disappearance of aesthetic nature from human environments means not only a reduction in quality of life, but a withdrawal of cultural possibilities. Addressing the interplay between culture and nature, Seel emphasises, for example, that aesthetic experiences of nature allow one to describe the quality of life in a certain environment and that this possibility establishes, among other things, their ethical value and relevance (Seel, 1996). Aesthetic nature is here not seen as a corporeal indicator; a beautiful landscape is not necessarily useful in a physical sense but as a special opportunity, a value that has ethical relevance not as food or living space, but for identifying possibilities of a felicitous human life in a eudaimonistic interpretation (Seel, 1996). Recently, eudaimonic values, aiming at higher levels of well-being as motivators for sustainable societal practices, have gained recognition in the sustainability debate (Neuteleers, 2020).

In summary, concepts of aesthetic experiences of nature and their values for humans are firmly embedded in socio-cultural and material contexts. The much-quoted "intrinsic value of nature" is, for example, conveyed culturally and symbolically and correlates philosophically with ideas of autonomy and a good, well-lived life.

Further, nature experiences and valuations are influenced by social differences. Smith (1984) emphasises, among others, that aesthetic experience of nature is mediated by socio-economic positions, access and historically produced spatial inequalities. Agyeman (2013) interprets nature experiences as socially unevenly distributed and as contributing to human well-being in ways that both reflect and reproduce social inequalities, thereby challenging universal claims of ethical values of nature.

In terms of environmental ethics, the intrinsic value of nature represents not a value per se, but a socially situated aesthetic and experiential value that emerges through differentiated material and cultural conditions. If intrinsic value is accepted as an aesthetic, experiential, and social (rather than ecological) value, it can inform environmental ethics discourses on just sustainable well-being, reconciling anthropocentric and biocentric interpretations and liberating the former from an overly instrumental perspective on nature.

For sustainable design acknowledging the autonomy of both nature and human beings means, in this line of argument, recognising that experiences of nature are culturally shaped. In practice SD as a cultural endeavour, grounded in the ethical and aesthetic dimensions of human–nature relationships, can include various indigenous interpretations of nature, such as viewing nature as a living relative or ancestral presence, which often interweave symbolism, collective memory and local environmental knowledge.

The theoretical discussion pauses here. The following chapter examines how the conceptual ideas introduced thus far unfold in diverse real-life settings, drawing on narratives from international design schools and their sustainable design curricula, as well as from intentional communities engaged in sustainable living.

Notes

1 For different ST levels, see European Environment Agency (2018). *Perspectives on transitions to sustainability*. Luxembourg: Publications Office of the European Union. https://www.eea.europa.eu/publications/perspectives-on-transitions-to-sustainability/file. Retrieved 30 March 2023.
2 The term "contractualism" denotes that morality is based on a contract or an agreement. Contractualism, Stanford Encyclopedia of Philosophy. Retrieved 30 March 2023. E.g., in Rawls: *"Thus, we are to imagine that those who engage in social cooperation choose together, in one joint act, the principles which are to assign basic rights and duties and to determine the division of social benefits"* (Rawls, 1971, 11).
3 Admittedly, this criticism also applies to Habermas's discourse ethics (DE), which nevertheless offers valuable direction for ethical decision-making. At its core, DE rests on the principles of autonomy, rationality, and communicative action presuming that human beings possess free will, the capacity for reason, and the ability to articulate justifications. According to Habermas (1990), universal norms can be critically evaluated and justified through emancipatory discourse, if communication unfolds within a hierarchy-free and egalitarian structure. Democracy, as an intrinsic political value, forms the civil foundation for such decision-making processes. Actors must offer valid justifications for their claims and actions. A "good reason" is not determined by one's social position or power, but by the intersubjective approval of others acting as rational respondents. Habermas' (1990) validity claims allow stakeholders to accept different methods of communication as possibilities for overcoming obstacles to achieve a common goal. This happens e.g. when at least two heterogeneous participants coordinate their perspectives and actions to succeed. Communication equals here coordination, and communicative acts focus on for example on goal-adjusted actions to integrate single performances in an action plan. DE allows a variety of views to be expressed and discussed. It is grounded in normative commitments to ethical values and civic virtues, including tolerance, respect for pluralism, and a willingness to seek compromise in pursuit of shared values and collective action (Keitsch, 2021). However, as a contractarian approach, DE faces challenges when confronted with real-world power asymmetries, ignorance, irrational behaviour, and individual self-interest which can severely impede realising emancipatory discourses.
4 According to Schatzki (2002), environmental actions are understandable only when viewed through *teleoaffective structures*, the interconnected goals and emotions that give practices meaning. Practices like water use or land management are shaped not just by policy or ecological facts, but also by what humans find meaningful and emotionally resonant within their specific cultural context. For example, indigenous environmental knowledge views nature as both a resource and a spiritual entity, generating values such as respect, reciprocity, and gratitude. Similarly, land conflicts often reflect competing teleoaffective structures: while one group may prioritise economic development, another may see the land as sacred. For Schatzki, these struggles are not merely political conflicts but are fundamentally rooted in different ways of seeing nature.
5 The terms "design for sustainable transitions" and "transition design" are used interchangeably.
6 According to Encyclopedia Britannica (n.d), situational ethics signify a view "that moral decision making is contextual or dependent on a set of circumstances … moral judgments must be made within the context of the entirety of a situation".
7 Societal actors are frequently categorised as either "stakeholders" or "users" (Reed, 2008). Stakeholders are individuals or groups who have an interest in a given situation and its development such as a proposed solution or who may be affected by it. In contrast, users are those who directly engage with or implement

the solution. Societal stakeholders include a wide range of actors, including national, provincial, and local governments; international organisations; research institutions and universities; public media and the press; non-governmental organisations (NGOs); private companies; grassroots organisations; and political activists. The goal of participatory processes is to design and implement strategies and mechanisms that enable both stakeholders and users to actively participate in, and contribute to, the decision-making, planning, and execution of projects and programmes.

8 Narrative ethics is an approach that emphasises understanding and interpreting individuals' stories as onset for ethical reflection, rather than relying solely on abstract principles or fixed rules in specific situations (Oxford Reference, https://www.oxfordreference.com/display/10.1093/oi/authority. Retrieved 30 November 2025). See also Wilks (2005). Fins et al. (1997) define clinical pragmatism as method for moral problem solving that comprises an analysis of intersubjective assessment and consensus development as well as moral considerations, which are illustrated through a detailed case study.Komesaroff (1995) introduced *micro ethics* as moral everyday decision-making. Micro ethics connect to relationships, intersubjective communication, and dialogues, and constitute the fundament for a generic ethical discussion. Even if not always based on conscious judgements, micro ethical decisions can be justified by virtue, which connects reflective and automatic choices. *Empirical ethics* is a counter-approach to so-called bioethical principlism as too abstract, universalistic, and rationalistic, proposing to support ethics with facts, experienced or more formally investigated (Parker, 2009). In case of participatory design this would mean to adjust values such as autonomy, justice and empowerment to the context in which they take place.

9 The historical debate on the relationship between aesthetic experience and participation can be traced back to the 3rd century to Plotinus, who imagined beauty as experience of the manifestation of God and His creation and aesthetic participation as partaking of this metaphysical totality. In the 18th century Idealist philosophers such as Kant (1790) promoted secularised contemplative participation as relationship between the aesthetic object and the autonomous subject – the beautiful "gives pleasure without interest" (*Critique of Judgment*, § 5,). Aesthetic participation as meditative (quasi) religious union appeared again in 1835 when Hegel discussed the artwork as representation of the world spirit, unembellished by any practical meaning or relevance to the daily life world (1975). As counter reaction to contemplative aesthetics, romanticism and naturalist art evolved in the 19th century. For naturalist art, especially poetry, the real human being as physical and social entity is the main topic of interest, as single individual, and as group in a socio-cultural context. In the 20th century, post-modernists, phenomenologists, critical theorists, and pragmatist approaches appeared, the latter, such as Dewey's aesthetics in 1934, emphasised the life-worldly, satisfying, didactic and political value of artworks (2005). All 20th-century aesthetic theories ascribe an important role to aesthetic participation in both social and individual development, albeit from heterogeneous philosophical positions (Keitsch, 2003b).

10 The modernism movement appears in the late 19th and early 20th century in philosophy, art and architecture with the aim to develop new representations that reflect the industrialising world, urbanisation, new products and processes, and, increasingly, political and environmental disturbances. Early modernism embraces the idea that scientific knowledge and technology will positively contribute to improve and reshape human surroundings, while modernism in the second decade of the 20th century thematises is more pessimistic discussing among others risks of science and technology, mass culture´s influence and human and societal commodification by capitalism (Butler, 2010).

11 The deductive-nomothetic method explains a phenomenon (*explanandum*) by deducing it from general laws and supporting statements (*explanans*). The

deductive aspect lies in the logical structure: the explanandum follows necessarily from the explanans. For example, the statement *"a tree is a living being"* (explanandum) can be derived from general traits of living organisms such as growth, having a lifespan, and eventually dying (explanans). The nomological aspect refers to the involvement of at least one law of nature as an indispensable premise. Without this law such as the biological principle that all living organisms have a lifespan the explanation would not hold. Scientific Explanation, *Stanford Encyclopedia of Philosophy.* (Retrieved 30 November 2023).

12 Critical rationalism with Popper (1957), argues that even if a statement cannot be logically deduced from existing knowledge, its falsifiability the possibility of being proven wrong contributes to the development of new knowledge. For example, the hypothesis *"All swans are white"* cannot be proven true, since not all swans can be observed. However, finding a black swan falsifies this claim and leads to a revised hypothesis, such as *"All swans are either white or black"*. This marks a clear departure from classical rationalism, which assumes that the truth or falsity of propositions can be determined independently of empirical observation and derived *a priori* through deductive reasoning. In contrast, empiricism relies on inductive reasoning, where knowledge is generated from observation and experience. Popper offers a synthesis: while analytic propositions (e.g., "all bodies have volume") are true by definition and not subject to falsification, synthetic propositions (e.g., "all tigers live in India") are based on empirical observation and cannot be conclusively verified, no matter how many confirming instances are observed. However, they can be falsified (e.g., by observing a tiger in China). A hypothesis that withstands multiple falsification attempts may be considered a tentative explanation, but never an absolute truth.

13 Foucault (1988b) describes four types of "Technologies of the Self": (1) technologies of production, which involve the creation and transformation of objects; (2) technologies of sign systems, which relate to communication through symbols and language; (3) technologies of power, which are used to influence or control behaviour; and (4) technologies of the self, which allow individuals to work on their own bodies, thoughts, and identities to reach particular goals.

14 A simple breakdown of Saussure's (1983) language model: The **signifier** is the form a sign takes. It can be a spoken word, a written word, a sound, or an image. For example, the word "tree" to see or to say is a signifier. It is the physical part of the sign.

The **signified** is the concept or idea of a signifier. Hearing or reading the word "tree" connects to think for example of a tall plant with branches and leaves (or to a friend). This mental image or idea relates to what is connected to the word.

A **sign** is created when the signifier and signified come together. The word "tree" (signifier) and the idea of a tree (signified) equal the sign. This combination gives meaning to language.

Signification is the process meaning making, that is when the signifier (the word or sound) and signified (the idea) are connected. Signification creates meaning, understanding and communication of language, symbols, and images.

Saussure emphasises that the relationship between a word (the *signifier*) and the idea it represents (the *signified*) is fundamentally arbitrary. There is no logical reason why the sound "tree" represents a particular object. It could have been any other sound "Baum", "Arbre", etc. what matters is that speakers of a language agree on its meaning. Meaning arises for Saussure from both social conventions and structural differences between signs within a language system. Supplementing this theory, Wittgenstein (1967) underlines the practical aspect of meaning, arguing its dynamic and context dependent character. Meaning is not fixed by mere definition but emerges and changes through use. The term *language-game* illustrates how words gain meaning through rule-governed, goal-oriented activities embedded in forms of life, making language is a living, dynamic practice.

15 The term "critical design" was coined by Dunne and Raby and first used in "Hertzian Tales" as part of Dunne's 1999 dissertation (Dunne, 2006). The boundaries between speculative design and critical design are fluid, and the two are collectively referred to in this book as speculative/critical design (SCD). SCD represents not a fixed concept, but an eclectic, fluid, and idiosyncratic disposition that opens space for a wide range of design strategies and conceptual explorations.

16 Early 19th-century romanticism emerged as a philosophical counterbalance to rationalism and empiricism, seeking the reconciliation of art, philosophy, and science. Drawing on the Renaissance ideal of a holistic culture, it positioned sensual experience, emotion, and feeling alongside reason. The romanticists also revitalised the notion of "genius", portraying the artist or philosopher as a messenger of higher truths. This resembles Plato's notion of *mania* in *Phaedrus* (1997, 245b). However, while Plato attributes creativity to divine madness, romanticism grounds it in subjective irrationality and emotion.

17 Critical theory, especially Horkheimer's interpretation of dialectics, has its roots in Hegel's philosophy. For Hegel, self-development occurs through a process called dialectical reflection, the progression of opposing ideas. It starts with a thesis (an idea), which is challenged by an antithesis (its opposite). The conflict between the two leads to a synthesis a resolution that combines elements of both. This synthesis then becomes a new thesis, and the process repeats. For example, the first idea is being (existence), which is challenged by its opposite, nothing. The synthesis is a new idea, becoming, which represents the process of change from nothing to being.

Adorno later reinterpreted this process as "negative dialectics" (2001) not to overcome contradictions, but to expose how many social structures are inherently false or unjust. For example, the culture industry produces mass entertainment that promotes conformity while creating the illusion of freedom via consumer choice. From a Hegelian perspective, the contradiction between conformity and freedom may resolve into a synthesis of progress or self-development. Yet, Adorno argued that what appears to be freedom such as choosing between entertainment options is actually a form of subtle control. Choices are pre-determined by the industry, leading consumers to conform without realising it.

18 Horkheimer interpreted critical theory as a heuristic tool: by exposing the mechanisms of domination (capitalism, ideology, instrumental reason), it could indirectly contribute to emancipation. Adorno was more pessimistic, claiming that a thoroughly commodified world makes true freedom or authenticity impossible: *"There is no right life in the wrong one". Minima Moralia* 18 (1945).

19 According to Ferrando (2013) posthumanism criticises anthropocentrism and reinterprets human relationality with technology, nature, and nonhuman others. Transhumanism discusses human enhancement through science and technology within a post- or hyper-biological future. Antihumanism deconstructs the subject. New materialism focuses on the agency of matter and the entanglement of human and nonhuman interactions.

20 The Speculative Turn, which evolved from the early 2000s onwards, challenges key assumptions of contemporary Continental philosophy, particularly the notion that thought and being are inseparable, and that humans can only access the world through language, consciousness, or subjective experience. In contrast, authors such as Meillassoux (2010) and Bryant et al. (2011) argue that reality exists independently of human perception, and that philosophy should engage directly with being itself, rather than restricting inquiry to human–world correlations. By advocating for realism and non-anthropocentrism, speculative philosophy seeks to move beyond the boundaries of human knowledge, aligning with the aims of posthumanism and new materialism, which both question human exceptionalism and foreground the agency of nonhuman entities.

21 Spinoza's Neoplatonic cosmotheology identifies God and Nature as one: *Deus sive Natura* ("God or Nature") rejecting any separation between the divine and the material world. God is immanent in all things, and everything in the cosmos is an expression of a single, infinite, and self-sustaining substance (*Ethics*, Part I). This view is influenced by Neoplatonism, which understands all of reality as originating from a unified, divine source (*Stanford Encyclopedia of Philosophy*, 2023).

22 Etiology means the study of causes, for example moral and social origins and effects.

23 "Chthonic – of or pertaining to the underworld". Latinised form of Greek *khthonios* "of the earth, in the earth". Online Etymology Dictionary (2023), https://www.etymonline.com/search?q=chthonic. Retrieved 30 November 2023.

24 Posthumanist concepts tend to overlook that there is no causality between ethical insight and ethical action, regardless of how profoundly entanglement with other life forms is experienced. Recognition of entanglement does necessarily lead to ethical behaviour or transformative action.

25 According to Merriam Webster (2024), "Liminal" signifes something that is barely perceptible or that involves a transitional or in-between state. https://www.merriam-webster.com/word-of-the-day/liminal-2021-08-22. Retrieved 30 November 2023. Characteristics of liminality are among others polyvalence, ambiguity, and temporal disorientation.

26 Regulative principles are maxims towards the improvement of concepts, such as environmental ethics or sustainable design. Constitutive principles are imperatives how concepts *must* be to are defined as ethical or environmentally sound (Blackburn, 2016).

27 Idiosyncratic knowledge refers to (inter)subjective, contextual understanding that comprises personal and collective perception, experiences, and values.

28 Rice (2018) outlines three mechanisms of human–nonhuman interaction in relation to artefacts: substitution, mediation, and communication. However, the examples Rice provides such as a speed bump substituting for a traffic warden to slow down motorists demonstrate that such tools are not deterministic, but contingent on social rules and regulatory frameworks. These artefacts may mediate human behaviour, but they do not constitute the norms themselves; instead, their function is shaped by the broader social systems in which they are embedded.

29 In 1935, Husserl coined the term "lifeworld" (Lebensumwelt, later Lebenswelt) describing immediate perceptions and experiences of the natural and social surroundings, which is distinguished for the objective world of science (although the latter has its origins in the former): *"the natural scientist does not realize that the constant foundation for his yet subjective reflections is the lifeworld, that is always assumed as the ground and field for work on which his questions and his methods alone have meaning*" (1935, Section II, translation M.K.)

German original: *"Aber der Naturforscher macht sich nicht klar, daß das ständige Fundament seiner doch subjektiven Denkarbeit die Lebensumwelt ist, sie ist ständig vorausgesetzt als Boden, als Arbeitsfeld, auf dem seine Fragen, seine Denkmethoden allein Sinn haben".*

30 Existential phenomenology focuses on an in-depth understanding of human existence, based on human embodiment and subjective experiences.

31 Due to translation difficulties, and when the term is unclear, I have chosen to add the German term.

32 Schön uses the term *talks back* to describe how professional practice prompts reflection, as practitioners engage in a dynamic dialogue with the situation at hand.: *"In the designer's conversation with the materials of his design, he can never make a move which has only the effects intended for it. His materials are continually talking back to him, causing him to apprehend unanticipated problems and potentials"* (2016, 100–101.)

33 *Heuristics* in the sense of Kant's "as-if" teleology: Kant viewed the purposiveness of nature as essential in the biological and context, which does not mean that intentions in nature can be proven; rather, teleology is necessary for human reflection and should therefore be imagined *as if* it was true. Kant's teleology can be assigned to the category of functional arguments; he himself presents it as heuristic by citing, within the framework of rational knowledge, that it is necessary to have this belief (Keitsch, 2003b): *"In order to see that a thing is only possible as an end, i.e. having to look for the causality of its origin not in the mechanism of nature, but in a cause whose ability to work is determined by concepts, requires: that its form is not possible according to mere natural laws, i.e. those which are determined by us the understanding alone, applied to objects of the senses, can be known; but that even their empirical knowledge, according to their cause and effect,* ***presupposes concepts of reason*** (emphasis M.K.). *This contingency of its form in all empirical natural laws in relation to reason, since reason, which must recognize the necessity of every form of a natural product if it only wants to understand the conditions associated with its production, but nevertheless cannot accept this necessity in that given form, is itself a reason to assume its causality* ***as if*** (emphasis M.K.) *it were only possible through reason; But this is then the ability to act according to purposes (a will); and the object, which is presented as possible only from this, would only be presented as an end for possible".* (Kant, *Critique of Judgement* § 64).

Intrinsicality refers in human–nature relationships to recognising nature's value beyond its utility for human purposes. It emphasises valuing nature for its own being, fostering a reciprocity where humans see themselves as participants within, rather than mere users of the natural world.

34 Ad. "thrownness" as reason for compassion. Heidegger sees temporality as primary condition of (human) being within the structure of existence, thrownness, and fallenness. *"Existence is the potentiality-for-being (Seinkönnen); it projects its being upon various possibilities. Existence represents thus the phenomenon of the future. Then, as thrownness, Dasein always finds itself already in a certain spiritual and material, historically conditioned environment; in short, in the world, in which the space of possibilities is always somehow limited"* (*The Internet Encyclopedia of Philosophy*, 2024, https://iep.utm.edu/heidegge/) From a NM perspective all natural beings would experience "thrownness" living in a certain ecosystem, which can evoke tthe compassion of being in a similar situation.

35 A category mistake means wrongly ascribing an attribute to something which belongs to another category, for example assigning an abstract quality to a physical being: "The cunning cobra looks for her prey". Or more non-sensical: "Anthropomorphism is eating its children".

36 "In der Tat zielt 'Big Nudging' auf die Gleichschaltung vieler individueller Handlungen und auf eine Manipulation von Sichtweisen und Entscheidungen ab. Dies rückt es in die Nähe der gezielten Entmündigung des Bürgers durch staatlich geplante Verhaltenssteuerung. Wir befürchten, dass die Auswirkungen langfristig fatal sein könnten". (Helbing et al. 2017, 18)

37 Vaihinger (1968) argues that empirically false assumptions can provide viable cognitive heuristics. Scientific models are for example often false formally, but they facilitate useful insights as idealisations. Similarly, teleological ideas are not true in an empirical sense, but they can facilitate moral reflections and actions.

38 In architecture, Norberg-Schulz (1926–2000) was an early representative of the place and atmosphere within a "genius loci" concept. Norberg-Schulz encourages the development of a poetic perspective towards "place", which allows room for interpretation of instrumental, physical, social and spiritual aspects: *"What then do we mean with the word 'place'? Obviously, we mean more than an abstract location. We mean a totality made up of concrete things, having material substance, shape texture and color. Together these things determine an 'environmental character'*

which is the essence of the place. In general, a place is given such essence or atmosphere" (1980, 6, 7).

39 In relation to the nature–art nexus, Adorno's reflections on affinity and polyvalence offer a valuable propaedeutic to environmental ethics. Affinity arises when nature appears to meet human expectations, generating a sense of resonance or familiarity. In contrast, polyvalence expressed through nature's indifference, destruction, and death evokes feelings of aporia (incomprehensibility) and amechanicha (helplessness). When confronted with such dissonance, practical responses whether moral, theoretical, or artistic often fail, particularly when the constructed image of human identity collapses under the weight of its own limitations. Yet, according to Adorno, no matter how intense the loss of control over thought and action, this paralysis is paradoxically accompanied by an experience of freedom the awareness that one can still act. Although humans are inevitably exposed to nature's polyvalence, autonomous artworks can evoke "negative freedom", pointing to insights that emerge through disorientation and rupture. *"An artwork opens its eyes under the gaze of the spectator when it emphatically articulates something objective, and this possibility of an objectivity that is not simply projected by the spectator is modeled on the expression of melancholy, or serenity, that can be found in nature when it is not seen as an object of action"* (Adorno, 1996, 275).

References

Adorno, T. W. (1993). *Beethoven: Philosophie der Musik*. Suhrkamp.

Adorno, T. W. (1997). *Aesthetic theory* (G. Adorno & R. Tiedemann, Eds.; R. Hullot-Kentor, Trans.). Minneapolis, MN: University of Minnesota Press.

Adorno, T. W. (2001). *Negative dialectics* (D. Redmond, Trans.; corrected ed. 2021). https://platypus1917.org/wp-content/uploads/Negative_Dialectics_Redmondtrans2021.pdf. Retrieved 28 December 2025.

Adorno, T. W. (2005). *Minima moralia: Reflections from damaged life* (E. F. N. Jephcott, Trans.). London: Verso.

Adorno, T. W., & Benjamin, W. (1999). *The complete correspondence, 1928–1940* (N. Walker, Ed.). Cambridge, MA: Harvard University Press.

Agyeman, J. (2013). *Introducing just sustainabilities: Policy, planning, and practice.* London: Zed Books.

Ajzen, I., Czasch, C., & Flood, M. (2009). From intentions to behavior: Implementation intention, commitment, and conscientiousness. *Journal of Applied Social Psychology*, 39, 1356–1372.

Akama, Y., Stuedahl, D., & Van Zyl, I. (2015). Design disruption in contested, contingent and contradictory future-making. *Interaction Design and Architecture(s)*, 26, 132–148.

Akenji, L., & Chen, H. A. (2016). *Framework for shaping sustainable lifestyles: Determinants and strategies.* Cambridge, MA: Harvard University Press.

Allmendinger, P., & Tewdwr-Jones, M. (2002a). The communicative turn in urban planning: Unravelling paradigmatic, imperialistic and moralistic dimensions. *Space and Polity*, 6(1), 5–24.

Allmendinger, P., Tewdwr-Jones, M., & Morphet, J. (2002b). Deconstructing communicative rationality: A critique of Habermasian collaborative planning. *Planning Theory*, 1(1), 9–26.

Araki, M. (2021). Four components of expertise (SSRN Scholarly Paper No. 3806101). Social Science Research Network. https://papers.ssrn.com/sol3/papers.cfm?abstract_id=3806101. Retrieved 24 March 2024.

Auger, J. (2013). Speculative design: Crafting the speculation. *Digital Creativity*, 24, 11–35.

Avelino, F. (2017). Power in sustainability transitions: Analysing power and (dis) empowerment in transformative change towards sustainability. *Environmental Policy and Governance*, 27, 505–520.

Barad, K. (2003). Posthuman performativity: Toward an understanding of how matter comes to matter. *Signs: Journal of Women in Culture and Society*, 28(3), 801–831.

Barad, K. (2007). *Meeting the universe halfway: Quantum physics and the entanglement of matter and meaning*. Durham, NC: Duke University Press.

Barad, K. (2010). Quantum entanglements and hauntological relations of inheritance. *Derrida Today*, 3(2), 240–268.

Bardzell, J., & Bardzell, S. (2013). What is critical about critical design? In *Proceedings of the SIGCHI Conference on Human Factors in Computing Systems*. https://www.researchgate.net/publication/262371319. Retrieved 14 April 2023.

Baldwin, C. (2015). Narrative ethics. In H. ten Have (Ed.), *Encyclopedia of global bioethics*. Cham, Switzerland: Springer.

Bennett, J. (2010). *Vibrant matter: A political ecology of things*. Duke University Press.

Berleant, A., & Carlson, A. (Eds.) (2007). *The aesthetics of human environments*. Peterborough, Canada: Broadview Press.

Berleant, A. (2013). What is aesthetic engagement? *Contemporary Aesthetics*, 11.

Berman, M. (1989). *Coming to our senses*. Simon & Schuster.

Blackburn, S. (2016). *The Oxford dictionary of philosophy*. Oxford University Press.

Blomkvist, J., & Holmlid, S. (2009). Exemplars in service design. In *Proceedings of the 1st Nordic Conference on Service Design and Service Innovation (ServDes 2009)* (pp. 1–10). Oslo School of Architecture and Design.

Bohman, J. (2005). Critical theory. In *Stanford Encyclopedia of Philosophy*. https://plato.stanford.edu/entries/critical-theory/. Retrieved 14 November 2023.

Böhme, G. (1993). Atmosphere as the fundamental concept of a new aesthetics. *Thesis Eleven*, 36, 113–126.

Böhme, G. (2010). The concept of body as the nature we ourselves are. *Journal of Speculative Philosophy*, 24(3), 224–238.

Böhme, G. (2017). *The aesthetics of atmospheres*. Routledge.

Bowen, S., McSeveny, K., Lockley, E., Wolstenholme, D., Cobb, M., & Dearden, A. (2013). How was it for you? Experiences of participatory design in the UK health service. *CoDesign*, 9(4), 230–246.

Braidotti, R. (2013). *The posthuman*. Cambridge: Polity Press.

Braidotti, R. (2016). Anthropos redux: A defence of monism in the Anthropocene epoch. *Frame*, 29, 29–46.

Bryant, L., Srnicek, N., & Harman, G. (Eds.) (2011). *The speculative turn: Continental materialism and realism*. Melbourne: re.press. https://re-press.org/book-files/OA_Version_Speculative_Turn_9780980668346.pdf. Retrieved 14 November 2023.

Buchmann, S. (2006). Conceptual art. In H. Butin (Ed.), *DuMonts Begriffslexikon zur zeitgenössischen Kunst*. Köln: DuMont.

Büsse, M. (2020). (Re)thinking design with new materialism. *Somatechnics*, 10, 355–373.

Butler, C. (2010). *Modernism: A very short introduction*. Oxford University Press.

Campbell, R. (1952). *Poems of Baudelaire*. Pantheon Books. https://fleursdumal.org/poem/214 (Retrieved 14 November 2023).

Carnap, R. (1931). Überwindung der Metaphysik durch logische Analyse der Sprache. *Erkenntnis*, 2, 219–241.

Cerpina, Z., & Stensli, S. (2022). *The Anthropocene cookbook: Recipes and opportunities for future catastrophes.* Cambridge, MA: MIT Press.

Chandler, D. (2019). *Semiotics for beginners.* https://www.cs.princeton.edu/~chazelle/courses/BIB/semio2.htm (Retrieved 20 February 2024).

Clavien, C. (2018). Ethics of nudges. *Journal of Moral Education*, 47(3), 366–382.

Clarke, B., & Rossini, M. (2017). Theoretical ecology and the posthuman. In B. Clarke & M. Rossini (Eds.), *The Cambridge companion to literature and the posthuman.* Cambridge, UK: Cambridge University Press.

Congiu, L., & Moscati, I. (2021). A review of nudges. *Journal of Economic Surveys*, 36, 188–213.

DeKay, M. (2011). *Integral sustainable design: Transformative perspectives.* Earthscan.

Derrida, J. (1978). *Writing and difference* (A. Bass, Trans.). University of Chicago Press.

Dewey, J. (2005). *Art as experience.* Tarcher Perigee (Penguin). Retrieved 20 February 2024.

Dieleman, H. (2008). Sustainability, art and reflexivity: Why artists and designers may become key change agents in sustainability. In S. Kagan & V. Kirchberg (Eds.), *Sustainability: A new frontier for the arts and cultures.* Verlag für Akademische Schriften.

Doyuran, E. (2023). Nudge goes to Silicon Valley: Designing for the disengaged and the irrational. *Journal of Cultural Economy.* https://www.tandfonline.com/doi/epdf/10.1080/17530350.2023.2261485. Retrieved 20 February 2024.

Dunne, A. (2006). *Hertzian tales: Electronic products, aesthetic experience, and critical design*. Cambridge, MA: MIT Press.

Dunne, A., & Raby, F. (2001). *Design noir: The secret life of electronic objects.* Springer Science & Business Media.

Dunne, A., & Raby, F. (2013). *Speculative everything: Design, fiction, and social dreaming*. Cambridge, MA: MIT Press.

Encyclopaedia Britannica (n.d.). Situation ethics. https://www.britannica.com/topic/situation-ethics. Retrieved 25 November 2025

Enzensberger, M. (1974). A critique of political ecology. *New Left Review*, 84, 3–31.

Eribon, D. (1992). *Michel Foucault*. Faber & Faber.

Ermacora, T., & Bullivant, L. (2016). *Recoded city: Co-creating urban futures.* Routledge.

Eronen, M. (2024). Aesthetic atmospheres and their affordances in urban squares. *Journal of Place Management and Development*, 17(3), 257–272.

Feola, G. (2020). Capitalism in sustainability transitions research: Time for a critical turn? *Environmental Innovation and Societal Transitions*, 35, 241–250.

Ferrando, F. (2013). Posthumanism, transhumanism, antihumanism, metahumanism, and new materialisms: Differences and relations. *Existenz*, 8. https://existenz.us (Retrieved 20 November 2023).

Fins, J. J., Bacchetta, M. D., & Miller, F. G. (1997). Clinical pragmatism: A method of moral problem solving. *Kennedy Institute of Ethics Journal*, 7(2), 129–145.

Fogg, B. J. (2009). A behavior model for persuasive design. In *Proceedings of the 4th international conference on persuasive technology* (Article No. 40).

Foucault, M. (1973). *Galskapens historie (The history of madness)*. Gyldendal.

Foucault, M. (1982). The subject and power. *Critical Inquiry*, 8(4), 777–795.

Foucault, M. (1984). Of other spaces: Utopias and heterotopias. *Architecture/Mouvement/Continuité*. https://web.mit.edu/allanmc/www/foucault1.pdf. Retrieved 20 November 2023.

Foucault, M. (1988a). *Schriften zur Literatur.* Fischer.

Foucault, M. (1988b). The political technology of individuals. In L. H. Martin, H. Gutman, & P. H. Hutton (Eds.), *Technologies of the self: A seminar with Michel Foucault* (pp. 16–49). University of Massachusetts Press. (Retrieved 20 November 2025).

Foucault, M. (2007). The incorporation of the hospital in modern technology. In J. Crampton & S. Elden (Eds.), *Space, knowledge and power: Foucault and geography* (pp. 141–152). Ashgate.

Friedland, R. (2018). What good is practice? Ontologies, teleologies and the problem of institution. *Management*, 21(4), 1357–1404.

Fry, T. (2009). *Design futuring: Sustainability, ethics, and new practice*. Berg.

Fung, A., & Wright, E. O. (2003). *Deepening democracy: Institutional innovations in empowered participatory governance*. Verso.

Gablik, S. (2003). Good citizenship: Design as a social and political force. In S. Heller & V. Vienne (Eds.), *Citizen designer: Perspectives on design responsibility*. Allworth Press.

Gagné, M., & Deci, E. L. (2005). Self-determination theory and work motivation. *Journal of Organizational Behavior*, 26(4), 331–362.

Gatto, G., & McCardle, J. (2019). Multispecies design and ethnographic practice. *Sustainability*, 11(18). n.p.

Gaziulusoy, A. İ., & Öztekin, E. (2018). Design as a catalyst for sustainability transitions. In C. Storni*et al.* (Eds.), *Design as a catalyst for change: DRS International Conference*. Limerick, Ireland.

Gaziulusoy, İ., & Öztekin, E. (2019). Design for sustainability transitions: Origins, attitudes and future directions. *Sustainability*, 11. DOI: doi:10.3390/su11133601.

Geels, F. (2005). *Technological transitions and system innovations: A co-evolutionary and socio-technical analysis*. Edward Elgar.

Gill, M. (2022). *A philosophy of beauty: Shaftesbury on nature, virtue, and art*. Cambridge University Press.

Gilligan, C. (2012). Can postmodernism contribute to saving the world? In R. McLean (Ed.), *Proceedings of the 11th European Conference on Research Methodology for Business and Management Studies* (pp. 461–469). Academic Publishing International.

Grin, J., Rotmans, J., & Schot, J. W. (2010). *Transitions to sustainable development: New directions in the study of long-term transformative change*. Routledge.

Habermas, J. (1990). *Moral consciousness and communicative action* (C. Lenhardt & S. W. Nicholsen, Trans.). Polity Press.

Halák, J. (2021). Revisiting Husserl's concept of *Leib* using Merleau-Ponty's ontology. *Southern Journal of Philosophy*, 59(3), 309–341.

Hamdi, N. (2010). *A place maker's guide to building*. Earthscan.

Haraway, D. J. (1997). *Modest_witness@second_millennium.femaleman©_meets_oncomouse*^{TM}*: Feminism and technoscience*. Routledge.

Haraway, D. J. (2007). *When species meet*. University of Minnesota Press.

Haraway, D. J. (2016a). *A cyborg manifesto: Science, technology, and socialist-feminism in the late twentieth century*. In *Manifestly Haraway*. University of Minnesota Press.

Haraway, D. J. (2016b). *Staying with the trouble: Making kin in the Chthulucene*. Duke University Press.

Hardt, M., & Negri, A. (2000). *Empire*. Harvard University Press.

Hartigan, J. (2019). Plants as ethnographic subjects. *Anthropology Today*, 35, p. 2.

Hargreaves, T. (2010). Putting Foucault to work on the environment: Exploring pro-environmental behaviour change as a form of discipline (*CSERGE Working Paper*

No. EDM 10–11). University of East Anglia. https://www.econstor.eu/bitstream/10419/48799/1/630850275.pdf. Retrieved 14 June 2023.

Harman, G. (2009). *Prince of networks: Bruno Latour and metaphysics.* re.press.

Hegel, G. W. F. (1975). *Aesthetics: Lectures on fine art* (T. M. Knox, Trans. Vols. 1–2). Clarendon Press.

Heidegger, M. (1971). Building dwelling thinking. In *Poetry, language, thought* (A. Hofstadter, Trans.). Harper Colophon Books. http://home.lu.lv/~ruben/Building%20Dwelling%20Thinking.htm. Retrieved 14 June 2023.

Heidegger, M. (1996). *Being and time* (J. Stambaugh, Trans.). SUNY Press.

Heidegger, M. (2002). *Off the beaten track* (J. Young & K. Haynes, Trans.). Cambridge University Press.

Heiskanen, E., Hyysalo, S., Jalas, M., Juntunen, J. K., & Lovio, R. (2014). User involvement and radical innovation: The case of heat pumps in Finland. In *Highways and byways of radical innovation: The perspective of design*. Kolding Design School.

Helbing, D. (2017). Big nudging – zur Problemlösung wenig geeignet. In C. Könneker (Ed.), *Unsere digitale Zukunft* (pp. 49–52). Springer.

Herbrechter, S. (2013). *Posthumanism: A critical analysis.* Academic Press.

Herrle, P., & Novy, J. (2012). *Social inclusion – A "new" paradigm for urban governance and development policy?*GIZ.

Hickey, S., & Mohan, G. (2004). Towards participation as transformation: Critical themes and challenges. In S. Hickey & G. Mohan (Eds.), *Participation: From tyranny to transformation?*Zed Books.

Hofman, L. (2023). Immanent obligations of response: Articulating everyday response-abilities through care. *Distinktion: Journal of Social Theory.* https://www.tandfonline.com/doi/full/10.1080/1600910X.2023.2222339. Retrieved 14 June 2023.

Hopkins, R. (2008). *The transition town handbook: From oil dependency to local resilience.* Chelsea Green Publishing.

Horkheimer, M., & Adorno, T. W. (1987). *Dialectic of enlightenment: Philosophical fragments.* https://monoskop.org/images/2/27/Horkheimer_Max_Adorno_Theodor_W_Dialectic_of_Enlightenment_Philosophical_Fragments.pdf. Retrieved 14 June 2023.

Huntjens, P. (2021). *Towards a natural social contract: Transformative social-ecological innovation for a sustainable, healthy and just society.* Springer.

Husserl, E. (1935). *Die Philosophie in der Krisis der europäischen Menschheit.* Wiener Kulturbund.

Huttunen, S., Ojanen, M., Ott, A., & Saarikoski, H. (2022). What about citizens? A literature review of citizen engagement in sustainability transitions research. *Energy Research & Social Science*, 91, 1–20.

Irwin, T. (2015). Transition design: A proposal for a new area of design practice, study, and research. *Design and Culture*, 7(2), 229–246.

Irwin, T., Tonkinwise, C., & Kossoff, G. (2015). Transition design: An educational framework for advancing the study and design of sustainable transitions. STRN Conference, University of Sussex. https://design.cmu.edu/sites/default/files/Transition_Design_Monograph_final.pdf (Retrieved 14 June 2023).

Iveson, R. (2015). Animals in looking-glass world: Fables of Überhumanism and posthumanism in Heidegger and Nietzsche. *Humanimalia*, 7(2).

Jameson, F. (1991). *Postmodernism, or the cultural logic of late capitalism.* Verso.

Johannessen, L., Keitsch, M., & Pettersen, I. (2019). Speculative and critical design: Features, methods, and practices. In *Proceedings of the 22nd International Conference on Engineering Design*. Cambridge University Press.

Johnson, D. W., & Johnson, F. P. (2009). *Joining together: Group theory and group skills* (10th ed.). Pearson.
Jones, S. (2024). Making time: Knitting as temporal-material entanglement. *Journal of Material Culture*, 29(1), 82–101.
Kagan, C., & Duggan, K. (2011). Creating community cohesion: The power of using innovative methods to facilitate engagement and genuine partnership. *Social Policy and Society*, 10(3), 343–354.
Kahneman, D. (2011). *Thinking, fast and slow*. Penguin Books.
Kant, I. (1788). *Critique of practical reason*. https://www.gutenberg.org/files/5683/5683-h/5683-h.htm (Retrieved 14 June 2023).
Kant, I. (1790). *Critique of judgment*. https://monoskop.org/images/7/77/Kant_Immanuel_Critique_of_Judgment_1987.pdf. Retrieved 14 June 2023.
Keitsch, M. (2003a). *Naturästhetik und ökologische Ethik: Eine Einführung*. Kovacs Verlag.
Keitsch, M. (2003b). Ästhetische Naturerfahrungen: "In-der-Welt-sein" ohne "außer sich" zu geraten. In N. Karafyllis (Ed.), *Biofakte*. Mentis Verlag.
Keitsch, M., Spencer, E., Sigurjonsson, J., & Nielsen, B. (2013). Sustainable design and social entrepreneurship: A common path toward social innovation? *Journal of US-China Public Administration*, 10(6), 618–629.
Keitsch, M. (2021). Enhancing collaboration between societal stakeholders for reduced inequalities. In *Encyclopedia of the UN Sustainable Development Goals*. Springer International.
Kemp, R., Weaver, P., Golland, A., Strasser, T., & Backhaus, J. (2018). *Sustainability transitions from a socio-economic transformation perspective*. European Environment Agency Report No. 25/2017, 70–94.
Kimura, H., & Nakajima, T. (2011). Designing persuasive applications to motivate sustainable behavior in collectivist cultures. *PsychNology Journal*, 9, 7–28.
Kipnis, A. (2015). Agency between humanism and posthumanism: Latour and his opponents. *Journal of Ethnographic Theory*, 5(2), 43–58.
Knox, J., & Pejan, R. (Eds.). (2018). *The human right to a healthy environment*. Cambridge University Press.
Knowles, B., Blair, L., Hazas, M., & Walker, S. (2016). Exploring sustainability through design fiction. In *Proceedings of the 2016 CHI conference on human factors in computing systems* (pp. 4034–4045). ACM.
Köhler, J., Geels, F., Kern, F., Markard, J., Onsongo, E., Wieczorek, A., Alkemade, F., & Avelino, F. (2019). An agenda for sustainability transitions research: State of the art and future directions. *Environmental Innovation and Societal Transitions*, 31, 1–32.
Koehrsen, J., & Huber, F. (2021). A field perspective on sustainability transitions: The case of religious organizations. *Environmental Innovation and Societal Transitions*, 40, 408–420.
Komesaroff, P. (1995). From bioethics to microethics: Ethical debate and clinical medicine. In P. Komesaroff (Ed.), *Troubled bodies: Critical perspectives on postmodernism, medical ethics, and the body*. Duke University Press.
Konior, B. (2021). We're all vermin: Tactical predation, interspecies media arts and perspectivism. *New Formations*, 104(1).
Kuhn, T. S. (1962). *The structure of scientific revolutions*. University of Chicago Press.
Kuruoğlu, A., & Woodward, I. (2021). Textures of diversity: Socio-material arrangements, atmosphere, and social inclusion in a multi-ethnic neighbourhood. *Journal of Sociology*, 57(1), 111–127.

Latour, B. (2005). *Reassembling the social: An introduction to actor-network theory.* Oxford University Press.

Leach, N. (1999). *The anaesthetics of architecture.* MIT Press.

Leach, N. (2016). Digital tool thinking: Object-oriented ontology versus new materialism. In *ACADIA 2016 Proceedings.* Posthuman Frontiers: Data, Designers, and Cognitive Machines. 344–351.

Leal Filho, W., Wolf, F., Lange Salvia, A., (2020). Heading towards an unsustainable world: Some of the implications of not achieving the SDGs. *Discover Sustainability,* 1, 2. https://link.springer.com/article/10.1007/s43621-020-00002-x (Retrieved 30 March 2023).

Lestar, T., & Böhm, S. (2020). Ecospirituality and sustainability transitions: Agency towards degrowth. *Religion, State and Society,* 48(1), 56–73.

Loh, J. (2022). Posthumanism and ethics. In: S. Herbrechter, I. Callus, M. Rossini, M. Grech, M. de Bruin-Molé, & C. John Müller (eds.) *Palgrave handbook of critical posthumanism.* Palgrave Macmillan.

Lohse, E., & Peters, J. (2023). *Sustainability through participation? National, supranational and international legal perspectives.* Brill.

Luederitz, C., Schäpke, N., Wiek, A., Lang, D., Bergmann, M., Bos, A., & Westley, F. (2017). Learning through evaluation: A tentative evaluative scheme for sustainability transition experiments. *Journal of Cleaner Production,* 169, 62–76.

Lyotard, J.-F. (1984). *The postmodern condition: A report on knowledge.* University of Minnesota Press.

Lyotard, J.-F. (1997). A postmodern fable. In *Postmodern fables* (G. Van den Abbeele, Trans.). University of Minnesota Press.

MacCormack, P. (2012). *Posthuman ethics: Embodiment and cultural theory.* Ashgate.

Magritte, R. (1929). *The treachery of images.* https://www.renemagritte.org/the-treachery-of-images.jsp. Retrieved 30 March 2023.

Malabou, C. (2022). *Plasticity: The promise of explosion* (T. Williams, Ed.). Edinburgh University Press.

Mander, J. (2012). *The capitalism papers: Fatal flaws of an obsolete system.* Counterpoint.

Marenko, B. (2015). *Deleuze and design.* Edinburgh University Press.

Marsh, D., & Stoker, G. (Eds.). (2010). *Theory and methods in political science* (3rd ed.). Palgrave Macmillan.

Mead, G. H. (1912). The mechanism of social consciousness. *The Journal of Philosophy, Psychology and Scientific Methods,* 9(15), 401–406.

Meillassoux, Q. (2010). *After finitude: An essay on the necessity of contingency.* Continuum.

Merleau-Ponty, M. (2002). *Phenomenology of perception.* Routledge.

Merriam-Webster Dictionary. (2024). Liminal. https://www.merriam-webster.com/dictionary/liminality. Retrieved 30 March 2024.

Mitrović, I. (2015). *An introduction to speculative design practice.* University of Split. https://www.researchgate.net/publication/276917770 (Retrieved 30 March 2023).

Mogstad, A., Høiseth, M., & Pettersen, I. (2018). Co-creation and co-design in public service innovation. In Proceedings of NordDesign 2018. Linköping, Sweden.

Mok, L., & Sampsa, H. (2018). Designing for energy transition through value sensitive design. *Design Studies,* 54, 162–183.

Morton, T. (2013). *Hyperobjects: Philosophy and ecology after the end of the world.* University of Minnesota Press.

Nelson, E. (2011). Revisiting the dialectic of environment: Nature as ideology and ethics in Adorno and the Frankfurt School. *Telos,* 155, 105–126.

Neuteleers, S. (2020). A fresh look at "relational" values in nature. *Environmental Values*, 29(4), 461–479.

Newell, P., & Mulvaney, D. (2013). The political economy of the just transition. *The Geographical Journal*, 179(2), 132–140.

Nielsen, A., Sand, H., Sørensen, P., Knutsson, M., Martinsson, P., Persson, E., & Wollbrant, C. (2016). Nudging and pro-environmental behaviour (TemaNord 2016: 553). Copenhagen, Denmark: Nordic Council of Ministers. https://norden.diva-portal.org/smash/get/diva2:1065958/FULLTEXT01.pdf. Retrieved 30 March 2023.

Nietzsche, F. (1873). On truth and lie in an extra-moral sense (W. Kaufmann, Trans.). https://jpcatholic.edu/NCUpdf/Nietzsche.pdf. Retrieved 30 March 2025.

Norberg-Schulz, C. (1980). *Genius loci: Towards a phenomenology of architecture.* Rizzoli.

Online Etymology Dictionary. (n.d.). https://www.etymonline.com. Retrieved 30 November 2023.

Opoku, H., & Keitsch, M. (2006). Une approche objective de la durabilité? Théorie des implications scientifiques et politiques de l'écologie industrielle. *Écologie & politique*, 32.

Oxford Reference. (n.d.). Narrative ethics. In *Oxford Reference.* Oxford University Press. https://www.oxfordreference.com/display/10.1093/oi/authority.20110803100223279. Retrieved 30 November 2025.

Parker, M. (2009). Two concepts of empirical ethics. *Bioethics*, 23, 202–213.

Parsons, G. (2008). *Aesthetics and nature.* Continuum.

Pavlov, I. P. (1928). *Conditioned reflexes.* Oxford University Press.

Pettersen, I., Geirbo, C., & Johnsrud, H. (2018). The tree as method: Co-creating with urban ecosystems. In *Proceedings of the 15th Participatory Design Conference* (Vol. 2, 1–6).

Pilemalm, S. (2018). Participatory design in emerging civic engagement initiatives in the new public sector. *ACM Transactions on Computer–Human Interaction*, 25, 1–26.

Plato. (1997). *Phaedrus.* https://xenopraxis.net/readings/plato_phaedrus.pdf. Retrieved 30 November 2025.

Plato. (2008). *Philebus.* Project Gutenberg. https://www.gutenberg.org/files/1744/1744-h/1744-h.htm (Retrieved 30 November 2023).

Popper, K. (1957). *The poverty of historicism.* Routledge.

Popper, K. (1996). *Realism and the aim of science.* Routledge.

Radomska, M. (2016). *Uncontainable life: A biophilosophy of bioart.* Linköping University Electronic Press.

Rawls, J. (1971). *A theory of justice.* Cambridge, MA: Harvard University Press.

Reed, M. S. (2008). Stakeholder participation for environmental management: A literature review. *Biological Conservation*, 141(10), 2417–2431.

Rekret, P. (2016). A critique of new materialism: Ethics and ontology. *Subjectivity*, 9.

Reno, M. (2023). Adorno on the possibility of nature. *Environmental Philosophy*, 20(1), 55–71.

Rice, L. (2018). Nonhumans in participatory design. *CoDesign*, 14, 238–257.

Riger, S. (1993). What's wrong with empowerment? *American Journal of Community Psychology*, 21, 279–292.

Sachs, J. D., Schmidt-Traub, G., Mazzucato, M., Messner, D., Nakicenovic, N., & Rockström, J. (2019). Six transformations to achieve the sustainable development goals. *Nature Sustainability*, 2, 805–814.

Sanders, E. B.-N., & Stappers, P. J. (2008). Co-creation and the new landscapes of design. *CoDesign*, 4(1), 5–18.
Sanders, E. B.-N., & Stappers, P. J. (2014). Probes, toolkits and prototypes. *CoDesign*, 10(1), 5–14.
Saussure, F.de. (1983). *Course in general linguistics*. Duckworth.
Schatzki, T. R. (2002). *The site of the social: A philosophical account of the constitution of social life and change*. Routledge.
Schmidt, A., & Engelen, B. (2020). The ethics of nudging: An overview. *Philosophy Compass*, 15(4).
Schön, D. A. (2016). *The reflective practitioner*. Routledge.
Seel, M., & Rigby, C. (1992). Aesthetic arguments in the ethics of nature. *Thesis Eleven*, 32(1), 76–89.
Seel, M. (1996). *Eine Ästhetik der Natur*. Suhrkamp.
Sharp, H. (2011). *Spinoza and the politics of renaturalization*. University of Chicago Press.
Shove, E., & Gordon, W. (2007). CAUTION! Transitions ahead. *Environment and Planning A*, 39(4), 763–770.
Simon, H. A. (1969). *The sciences of the artificial*. Cambridge, MA: MIT Press.
Singh, S., & Keitsch, M. (2018). Aesthetic participation for sustainable development. *Journal of Management and Sustainability*, 8(4), 96–103.
Skinner, B. F. (1973). *Beyond freedom and dignity*. Penguin Books.
Smith, K. (2009). *This is not a book*. Penguin.
Smith, N. (1984). *Uneven development: Nature, capital, and the production of space*. Blackwell.
Soper, K. (2012). The humanism in posthumanism. *Comparative Critical Studies*, 9(3), 365–378.
Sousa, L.de. (2018). The lived body as pre-reflective consciousness. *Studia Universitatis Babe⊠-Bolyai Philosophia*, 63, 7–20.
Spahn, A. (2016). The first generation to end poverty and the last to save the planet? *Sustainability*, 10, 2–16.
Spinoza, B. (2009). *Ethics*. Project Gutenberg. https://www.gutenberg.org/ebooks/3800 (Retrieved 30 November 2023).
Stanford Encyclopedia of Philosophy. (2023). Neoplatonism. https://plato.stanford.edu/entries/neoplatonism/. Retrieved 30 November 2023.
Sterling, S. (2003). *Whole systems thinking as a basis for paradigm change in education*. University of Bath. http://www.bath.ac.uk/cree/sterling/sterlingthesis.pdf. Retrieved 30 November 2023.
Sumartojo, S., & Pink, S. (2019). *Atmospheres and the experiential world: Theory and methods*. Routledge.
Sundberg, J. (2014). Decolonizing posthumanist geographies. *Cultural Geographies*, 21(1), 33–47.
Sunstein, C. R. (2015). The ethics of nudging. *Yale Journal of Regulation*, 32, 413–450.
Sunstein, C. R. (2016). People prefer System 2 nudges (kind of). *Duke Law Journal*, 66, 121–168.
Swann, C. (2002). Action research and the practice of design. *Design Issues*, 18(1), 49–61.
Świątek, M., Szcześniak, H., Borkowska, W., Myszak, B., & Myszak, U. (2023). Aesthetic experience and the ability to integrate beauty. *Psychology Research and Behavior Management*, 4033–4041.
Tagliabue, M., & Sandaker, I. (2019). Societal well-being. *Behavior and Social Issues*, 28, 99–113.

Tate. (2025). *Fat chair* (Joseph Beuys, 1964–1985). https://www.tate.org.uk. Retrieved 1 January 2025.
Thaler, R. H., & Sunstein, C. R. (2008). *Nudge: Improving decisions about health, wealth, and happiness.* Yale University Press.
The Internet Encyclopedia of Philosophy. (2024). Heidegger. https://iep.utm.edu/heidegge. Retrieved 1 January 2024.
Treaty on European Union. (2023). Principle of proportionality (Article 5(4)). https://eur-lex.europa.eu/EN/legal-content/glossary/principle-of-proportionality.html. Retrieved 30 November 2023.
Ulmer, J. (2015). Plasticity: A new materialist approach to policy and methodology. *Educational Philosophy and Theory*, 47(10).
United Nations Environment Programme. (2016). *A framework for shaping sustainable lifestyles: Determinants and strategies.* https://www.oneplanetnetwork.org/sites/default/files/a_framework_for_shaping_sustainable_lifestyles_determinants_and_strategies_0.pdf. Retrieved 30 November 2024.
Vaihinger, H. (1968). *The philosophy of "as if".* Routledge & Kegan Paul.
Van Assche, K., Duineveld, M., Beunen, R., Valentinov, V., & Gruezmacher, M. (2022). Material dependencies. *Journal of Environmental Policy & Planning*, 24 (3), 281–296.
Van Gogh Museum. (2024). *Shoes.* https://www.vangoghmuseum.nl/en/collection/s0011V1962. Retrieved 30 November 2023.
Veslova, E. (2019). Design for sustainable entangled human–nature systems. In NORDES 2019: Who cares?Espoo, Finland. https://dl.designresearchsociety.org/cgi/viewcontent.cgi?article=1540&context=nordes. Retrieved 30 November 2023.
Visser, F., Stappers, P. J., van der Lugt, R., & Sanders, E. B.-N. (2005). Contextmapping. *CoDesign*, 1(2), 119–149.
Wang, P., & Yu, C. (2018). Aesthetic experience as an essential factor. *Sustainability*, 10, 1098.
Wang, X., & Lo, K. (2021). Just transition: A conceptual review. *Energy Research & Social Science*, 82.
Ward, K. (2020). Breaking down experience. *European Journal of Philosophy*, 29(4), 712–730.
Ward, M. (2019). *Thoughts on critics of critical and speculative design.* https://speculativeedu.eu/critical-about-critical-and-speculative-design/. Retrieved 30 November 2023.
Welsch, W. (1987). *Aisthesis.* Klett-Cotta.
West-Eberhard, M. J. (2003). *Developmental plasticity and evolution.* Oxford University Press.
White, D. (2020). Just transitions/design for transitions. *Capitalism Nature Socialism*, 31(2), 20–39.
Wilks, T. (2005). Social work and narrative ethics. *British Journal of Social Work*, 35(8), 1249–1264.
Winner, L. (1993). Opening the black box and finding it empty. *Science, Technology & Human Values*, 18(3), 362–378.
Wittgenstein, L. (1967). *Philosophical investigations.* Blackwell.
Whyte, K. (2018). Indigenous science (fiction) for the Anthropocene. *Environment & Planning E: Nature and Space*, 1(1–2), 224–242.
World Commission on Environment and Development. (1987). *Our common future.* Oxford University Press. https://sustainabledevelopment.un.org/content/documents/5987our-common-future.pdf. Retrieved 30 November 2024.

Zamenopoulos, T., & Alexiou, K. (2018). *Co-design as collaborative research.* Bristol University/AHRC Connected Communities Programme. https://oro.open.ac.uk/58301/1/Co-Design_CCFoundationSeries_PUBLISHED.pdf. Retrieved 10 February 2024.

Žižek, S. (2014). *Absolute recoil: Towards a new foundation of dialectical materialism.* Verso.

4 International Sustainable Design Practice

4.1 Exploring Environmental Ethics and Sustainable Design through Narratives

The first three chapters of this book discussed environmental ethics and sustainable design through concepts from various authors. Chapter 4 differs in its methodological approach. It employs narratives to illustrate examples drawn from sustainable design curricula and sustainable living initiatives. The purpose of these narratives is to highlight significant idiosyncratic features of individual cases that revolve around the themes discussed in the previous chapters, namely human–nature relationships, interpretations of human (professional) responsibilities, and views on the role of products, services, and systems.

Narratives present perspectives, values, norms, and actions in the form of storytelling (Hughes et al., 2002)[1]. They invite readers to move beyond disciplinary epistemologies, encouraging them to reflect, develop, or transform their views. Narratives are used across various disciplines. In philosophy, for example, narratives convey reflections on human existence from antiquity to modernity, as in Schweitzer's nature reflections:

> *"On the river, surrounded by the wild splendor of nature, I became aware of something mysterious within all things, something which spoke to me of life itself. The water rippled by with a murmuring sound that seemed to tell of endless striving. The trees on the banks stood silent, yet full of life, reaching their branches to the sky as if in prayer. Birds flew overhead, calling to each other, and tiny insects hovered and darted in the air. I felt myself united with them, sharing their will to live, their quiet joy in existence. Then it became clear to me, as by an inner revelation: I am life that wills to live, in the midst of life that wills to live. From that moment on, I knew my task was to show reverence for all life, to act in ways that preserve and honor it wherever I could".*
>
> (Schweitzer, 1933, 116–117).[2]

In design, narratives can exert a persuasive influence by telling stories of how users and stakeholders successfully engage with products, services, or systems. Designers often use such narratives to entice customers or to create products

DOI: 10.4324/9781003384090-4

with greater user appeal. By contrast, emancipatory narratives suggest that design can transcend mere product development to foster empathy, critical reflexivity, and transformation. Fry (2022), for example, argues that emancipatory narratives emphasise the designer's role in facilitating deeper social change and personal growth.

Before discussing the advantages, challenges, and practical aspects of using narratives, a brief remark on their philosophical underpinnings is needed. A naturalist view holds that events and phenomena precede any narrative, while a social constructivist view claims that the narrative precedes the events. The narrative approach in this book seeks to reconcile this antinomy. Ontologically, the author assumes an intersection of the sequential nature of events (empirically), their embeddedness in experience (phenomenologically), and their (socially constructed) discursive structure. The methodological rationale for using narratives is to connect nomothetic with idiographic knowledge.

The term *idiographic* stems from the Greek *idios* (own, personal, individual) and *graph-* (root) (Word Reference.Com, 2024). Windelband introduced the term in 1894 to distinguish between the Humanities and the Natural Sciences.

> *"the strength of natural science lies on the side of abstraction, that of history on the side of intuitive concreteness ... through the mouth of history, past languages and past peoples speak to us, risen from forgetfulness into new life their beliefs and forms, their striving for power and freedom, their poetry and their thought. How different is the world that natural science builds before us! However vivid its points of departure may be, its cognitive goals are theories ... it leaves behind the individual sensory thing that comes into being and passes away as a semblance without essence, and strives toward knowledge of the lawful necessities that reign over all events in timeless immutability ... Indifferent to the transient, it casts its anchors into that which eternally remains identical with itself; it does not seek the changeable as such, but the unchanging form of change".*
>
> (ibid., trans. M.K.)[3]

Nomothetic knowledge seeks to generalise information with universal validity, viewing individuals as examples of general patterns, for instance when assessing data to identify trends in human behaviour. Idiographic knowledge, by contrast, seeks validity in unique individuals or phenomena, aiming to understand general features through specific cases. Qualitative methods, such as narratives, are employed to interpret subjective experiences and situational contexts, which can significantly deepen the understanding of a case. Windelband underscores, however, that the distinction between nomothetic and idiographic knowledge is merely analytical: any research object may contain elements of both, depending on the purpose of the investigation[4].

Idiographic knowledge generation is also central to phenomenological inquiry, which aims to explore individual experiences and interpretations. Van Manen et al. (2021) note that the strength of a phenomenological method lies in its in-depth description of a specific situation, while its validity rests in achieving an *"universal singularity"* (ibid.,1079).

The narratives in this book represent a scientific-literary hybridity that offers a way to understand theory–practice relations in environmental ethics and sustainable design. They draw inspiration from a performative approach by integrating a literary form into an analytic research process. Performativity broadens representational possibilities by blending artistic performances with science, enriching logical, nomothetic explanations with individual, sensuous, and multifaceted meanings (Cattermole et al., 2021). According to Gergen and Gergen (2010), the intersection of science and art has a long history; theories often incorporate literary tropes such as narratives and metaphors, while photographs, animations, and illustrations are frequently used to clarify scientific data and models. Recent authors (Heras et al., 2014; Brossmann et al., 2020; Cattermole et al., 2021) also highlight how performative approaches can deepen designers' understanding of sustainability. By connecting data and experimental findings with experiences, needs, and human values, such approaches can promote proactive engagement with sustainability. Performative research is explored further in the subchapter 5.3.1.

The narratives in this book do not, however, adhere strictly to a performative research paradigm (Østern et al., 2023) but are situated within a qualitative research framework. Their content is based on case observations, semi-structured interviews, and document analysis, expressed through the author's interpretation of real-world situations in narrative form. Hughes et al. (2004) emphasise that the internal validity of a narrative in social research depends on its purpose and significance for the overall study. Narratives do not represent facts; rather, they enrich the interpretation of settings, making them highly relevant for design. They are also recognised as a valuable qualitative research method in social research (Hughes et al., 2004).

On the downside, a "make-sense approach" (Perkins et al., 1991, p. 99) can affect the reliability of findings. Personal values, experiences, and philosophical perspectives such as the aim of identifying how theories manifest in practice and uncovering underlying themes can influence case selection and interpretation. Another limitation is confirmation bias, which can be particularly strong in cultural settings that differ significantly from the author's previous research contexts. To mitigate this, the author engaged in critical post-reflection, complemented as far as possible by feedback from interviewees and by discussions of the narratives with fellow researchers.

The narratives presented in this chapter are interpretive constructions to illustrate both the specificity of particular contexts and their relevance to the broader themes developed earlier in the book. Rather than functioning as conventional case studies, they are assembled from a diverse range of empirical materials, including stakeholder interviews, interviews with educators and students affiliated with different design schools and intentional communities, observations, document analysis, and the author's own experiential insights. Data collection was conducted between 2023 and 2025 through repeated site visits of varying duration.

To ensure anonymity, all names, locations, and identifying details have been modified. The personae presented in the narratives are composite constructs derived from recurring perspectives and discussions across the interviews. To the greatest extent possible, interviewees were given the opportunity to comment on and respond to the narratives, and their feedback has been incorporated to enhance contextual grounding and interpretive accuracy. The narratives should be read as reflexive interpretations that integrate empirical material with the author's theoretical reflections, rather than as verbatim accounts.

4.2 Graduate School of Design and Intentional Sustainable Community, Boston, United States of America

4.2.1 Design as Fairness

This is an Ivy League institution, where historic buildings convey tradition, status, and affluence. In contrast, the Graduate School of Design is a modern concrete construction. The institution is not only renowned for academic excellence but also serves as a significant hub of political thought influencing the country's broader political landscape. Left-wing and right-wing ideologies are present across different schools, with a prevalence of liberal thoughts. Reconciling progressivism and conservatism is challenging and highly political at the institution. This reflects the broader political discourse in the United States, as well as the cultural and social dynamics in the country.

Discussions on the realisation of societal and sustainability values, alongside the ambition for world-renowned scholarship and professionalism, are an integral part of education and research at the school. Yet, navigating differing values can be challenging, and students may feel uncertain about working with clients or groups whose perspectives differ from their own. The curriculum is hence designed to cultivate professionalism rooted in critical reflection rather than shaped by personal bias, and a central aim is to help students engage with the interplay between diverse stakeholder perspectives, design activism, and professional ethics maintaining a client-centred approach while pursuing practical, context-sensitive solutions.

One way to integrate design professionalism with value-based standpoints is through examples and case studies presented by design practitioners. Helen, a faculty member, emphasises that experiential engagement with real-world cases constitutes a central pedagogical approach in design education. Focused on learning-by-doing, ten to twelve design experts are invited each semester to the design master's course to share insights on their professional careers. Besides developing distinctive design skills, the students experience here how design practice is dealt with by different experts. The practitioners, who are alumni from GSD and other schools of the university, have very distinct career paths. They represent a wide selection of expertise and career stages, ranging from 2 to 20 years of experience, including engineers, educators, activists, NGO members, and government officials working on projects related to sustainable development goals, urban planning, sustainability, and social inclusion. They comprise young professionals just starting their careers, people working specifically with sustainability design and planning, and senior professional planners, among others.

Helen's goal is to provide a broad overview of professionals and how they address social and sustainability issues. It is particularly important that the students gain an understanding of working in diverse societal, environmental, and geographic contexts and realise that value conflicts in projects are normal rather than exceptional. Navigating value conflicts and professionally developing satisfying solutions for clients, stakeholders, or communities do not exclude each other but are part of a designer's and planner's professional life. The professional autobiographies illustrate ways, possibilities and also challenges of managing value conflicts and how to achieve a fair balance of both political and professional values. Sometimes, diverse stakeholders in a project can establish values, goals, and actions that allow working towards a broader common objective. Achieving healthy environments that enhance the residents' quality of life is, for instance, a shared objective, which can transcend political opinions and individual inclinations. The designer's role is to facilitate collective understanding and goal setting, rather than impose individual opinions or create division. Political and professional practices must be harmonised to effectively address social challenges, and the biographies encourage students to develop humility regarding their own ideologies. They train them to engage in emancipatory discourses and to detect common opportunities through listening to stakeholders, especially those with differing perspectives. The students learn to understand the actual needs of communities and learn to decide when to emphasise and when to retreat from their professional stance. They also gain practical knowledge on how to emphasise their values sensibly, which is a profound learning experience.

For Helen, it is crucial for the students' professional development that the experts present real-life cases, connecting academic knowledge with practical expertise and highlighting possibilities and challenges regarding the dilemma above. Besides fostering students' critical thinking, another advantage of integrating professionals into sustainable design education and planning is that these practitioners often work with cutting-edge environmental technology and develop and implement innovative sustainable design solutions. Their frequent transdisciplinary interaction across various fields enables students to recognise the interconnectedness of sustainability issues and the systemic impact of potential design solutions. Challenges for including practitioners in the curriculum comprise, among others, lacking pedagogic skills, the diversity of tools and methods potentially confusing students, biases toward design and planning approaches, and intellectual property issues, for instance when mediating methods or case studies from professional work. Balancing political assertiveness, design agency, and professionalism, practitioners are, however, priceless role models and mentors who provide expertise and inspiration. Their guidance and inspiration are immensely valuable, benefiting students, educators, practitioners, and clients alike in the field of sustainable design.

My impression is that the school places strong emphasis on issues such as equity, access, health, and climate. Within this context, social sustainability particularly as it relates to community living and quality of life appears to be prioritised over more philosophical discussions concerning the role of nature. The interviews also foregrounded reflections on the relationship between theory and practice in design, as well as on the ethical responsibilities of designers in relation to sustainability and social justice.

The following narrative is informed by conversations with a senior faculty member in urban design and planning whose work draws on extensive experience in the study of urbanism across diverse global contexts. Rather than offering a descriptive account of curricular content, the narrative adopts sustainable design fiction as a methodological approach to more effectively convey the substance of these discussions. By deliberately combining empirical insights with speculative elements, the narrative challenges prevailing assumptions and articulates a vision of a socially and environmentally progressive future.

4.2.2 The Aghori Agency[5] *– A Design Fiction*

Mumbai 2084: a metropolis of 38 million inhabitants, with almost half its population living in informal settlements. The Aghori community, a former esoteric sect dedicated to individual liberation, has transformed into a broader social movement in India, promoting justice and inclusion across the country. The core value of the Aghori, formerly centred on attaining personal liberation, is today supplemented by the commitment to work for the spiritual development and welfare of all living beings. With deep compassion for the socially underprivileged, the Aghori movement particularly combats environmental and social devastation in India's megacities.

Some planners and urban experts claim that Mumbai's squatter settlements represent the demise of modern city utopias. However, professionals who embrace Aghori philosophy view informal settlements as both challenges and opportunities. Contrasting the planned and bureaucratised parts of the city, which are characterised by excessive high-rise residential buildings designed to accommodate the increasing urban population, informal settlements embody a dynamic fabric that disrupts static structures. While extreme urban environmental problems prevail in the city, some of Mumbai's squatter settlements have emerged as laboratories for initiating change for a wider population through environmentally and socially conscious interventions.

The informal settlement of Shivaji Nagar belongs to the most impoverished and marginalised urban areas in Mumbai, housing approximately 1.9 million residents. Poor Muslims, migrants, and Dalits live crowded into 1.97 square kilometres, next to the Deonar landfill. The settlement suffers from a severe lack of health facilities. There is only one toilet for every 145 people, and during the monsoon season, homes frequently endure sewage overflow from communal pit latrines. Inhabitants without proper dwellings must cook and defecate in front of the small plastic sheets that serve as their shelters. The constant threat of contamination of drinking, cooking, and washing water by excrement and other pollutants is a persistent worry for most residents. Shivaji Nagar is frequently devastated by cholera and other excrement-associated diseases.

In 2080, a group of Brahmins[6], inspired by the Aghori philosophy, including engineers, planners, and investors, decided to transcend the ingrained Hindu doctrine of impurity. Together with Shivaji Nagar residents and local representatives, they embarked on planning, designing, and implementing revolutionary sanitation concepts in the settlement. Some of the group members stayed in the settlement to experience the situation firsthand. Shraddha Sharma, a Brahmani, architect, and supporter of the Aghori movement, explains: *"Besides the sickening state of the toilets and the lack of doors, safety was a major concern. As a woman, I constantly feared being attacked or molested. To avoid using the toilets, I stopped drinking water, which resulted in constipation. Using the toilets during menstruation was a real nightmare".*

The true radicality of the sanitation project lay not in its innovative and participatory environmental technology, but in the revolutionary mental and social shift it symbolised. Brahmins, who are not supposed to touch both casteless individuals and any form of human waste, continuously encountered pollutants, and the project challenged deeply ingrained religious taboos and norms and triggered various heated debates over banning caste discrimination.

The experts and investors were guided by two interrelated principles: sanitation is intrinsically linked to equity, and to understand sustainable sanitation, we must expand our knowledge to encompass all available information and experience and adapt technology to the needs of the setting. The project embraced transdisciplinary collaboration, including a wide group of stakeholders from Shivaji Nagar. The project initiators possessed the political and economic power necessary for implementation, along with technical and planning expertise, while the stakeholders contributed practical skills and knowledge on conditions and circumstances in the settlement. Following phases of identifying, discussing, and testing possibilities in a designated area of Shivaji Nagar, one sanitation concept and one supplementary solution were implemented in 2083 across the entire settlement, aiming to transform waste into a valuable resource by creating a self-sustaining, income-generating system that is maintained and capitalised by the community. The core concept comprises energy-producing toilets that generate electricity from human waste via microbial fuel cells, which convert organic waste into electricity. To optimise space usage, eighty vertical sanitation towers were constructed in small available spaces. Each of the Satchhata towers houses 20 separate toilets for men and women on two floors, washing facilities, menstruation products, and water purification systems. The modular units incorporate greywater recycling to reuse water, and each tower has a rooftop electricity generation system, where batteries store the energy generated.

The design also comprises two green walls with fluorescent hydroponics to support water purification, improve air quality, and enhance aesthetics. Maintenance of the buildings, disease outbreak monitoring, and distribution of electricity are organised by the community. The Shivaji Nagar sanitation committee exploits blockchain technology to meet these tasks. Blockchain-based voting ensures resident participation in decision-making. Additionally, the committee issues tokens for community goods, services, or utilities as incentives for the proper use of sanitation facilities. The users register with their fingerprint and get a PIN code for access to the male or female toilet section. Excess electricity is sold to Mumbai's energy companies. The second concept addresses toilet cleaning, responding to the fact that it provides an essential public service, but often at the cost of the sanitation workers' dignity, safety, and health. The toilets in the Satchhata towers are cleaned thrice daily by robots that can identify dirt and stains, register the refill of soap and menstruation products, and have a self-cleaning and self-emptying system to dispose of waste. The former Bhangis[7] are trained and employed to manage storage, operation, maintenance, and troubleshooting of the cleaning robots. They also work as ambassadors informing on proper toilet use and encouraging users to give feedback on cleanliness. They conduct a three-monthly evaluation of the robots' efficiency and suggest improvements for the robots' operation and the overall toilet maintenance.

The Satchhata towers have been operational for a year, from 2083 to 2084, a too brief period to evaluate their long-term health and environmental benefits for the Shivaji Nagar population. Surveys show, however, that contentment with the sanitation situation in the area has increased and that the electricity generation has opened new business opportunities in the settlement. The towers also contribute to social cohesion by establishing a sanitation committee and resident engagement, and the tokens motivate proper use of the facilities. The towers exemplify a promising approach for other informal settlement areas, and they are adaptable to different environments and community preferences. Ragnesh Kumar Pandey, the overall project lead, acknowledges the profoundly provocative aspects associated with the Satchhata towers, yet he firmly believes that solutions like these immensely enhance the quality of life in informal settlements and can serve as a blueprint for sustainable sanitation under extreme conditions. Moreover, they embody a crucial shift in professionals' mindset toward social equality: *"Professionals are agents of practice. Bearing critical responsibility and moral reflection on the space and its possibilities is a pivotal task of an architect".*

4.2.3 Building a Community Before a Community Building

The following narrative draws on interviews and observations conducted in two intentional community settings and examines key dimensions of social sustainability as articulated and enacted in everyday life by the inhabitants.

AUTHOR: I'm here in the lounge and common space of the Co-Housing Community in Boston, speaking with Avery, Jordan, Casey, who are long-standing members of this co-housing project. Thank you all for taking the time to talk with me today. To begin, could you share what initially inspired the creation of the Co-housing project?

AVERY: We wanted to get to know each other before we started planning and living together. Co-housing is a collaborative project for co-living, sharing space, and practising sustainability. The idea was to establish a shared vision as a foundation for the project. Before we even began looking for land, we invested time in collective discussions to clarify our goals and expectations.

JORDAN: Exactly. Co-housing is part of a broader cohousing movement across the US, which seeks to advance social relations, environmental responsibility, and inclusivity. Currently, there are approximately 175 co-housing communities in the country. Like many of them, ours combines private housing with shared spaces designed to support both individual autonomy and collective well-being.

CASEY: Construction of our co-housing complex started in 2020, but the foundations of the community were established as early as 2016, when we first came together. Having known one another for an extended period helped to build trust and facilitated more effective collective decision-making.

AVERY: In the early stages, we met on a monthly basis to discuss our plans and align our expectations. A key reference for us was Chris Scott-Hanson's *Cohousing Handbook*, which offers structured guidance for establishing co-housing communities. We followed its framework closely, and it provided a shared point of reference throughout the process.

AUTHOR: So, what would you identify as the key elements that contributed to the project's success?

CASEY: We shared common values and long-term goals from the outset. Over time, individuals with diverse professional backgrounds such as academics, lawyers, engineers, and social scientists joined the group, some retired and others still professionally active. This broadened the group's collective expertise and proved valuable in addressing practical and organisational challenges. We also became members of the homeowners' association early on, which allowed us to act collectively within local land-use regulations.

JORDAN: Finding a suitable location was initially challenging. We considered acquiring land outside of Boston to pursue permaculture and organic farming, as plots were more affordable and readily available. Ultimately, however, we selected a site near the Riverside District. The next phase involved collaboration with several architects, after which an international architecture practice was commissioned. Their brief was to design a building that would integrate into a low-density neighbourhood while supporting both private living and communal activities. The completed building includes individual residential units connected through shared spaces, terraces, a dining hall for up to 100 people, a music room, meeting rooms, and a community pantry. These spaces support both privacy and collective activities such as childcare and yoga. The site covers approximately 3,000 square metres, and careful attention to light and spatial organisation contributes to a sense of openness.

CASEY: We placed particular emphasis on creating accessible shared spaces to ensure that residents with disabilities could participate fully in community life. One feature we are especially proud of is the open hallway design, which allows residents with limited mobility to engage with their neighbours from their windows.

JORDAN: Social interaction was a central value guiding the building's design. We wanted the architecture to encourage care and mutual support among residents. Green spaces were also prioritised, including a small resident-managed garden, as well as generous greenery surrounding the apartments. The site is adjacent to a river corridor with a walkway that residents can access and enjoy.

AVERY: The project was completed in 2022. All units have since been sold, and demand for the apartments remains high. We do regret that resident diversity is more limited than we had hoped. Housing prices are high so this presents a significant barrier. We attempted to increase diversity by renting two units to socially disadvantaged families, but this initiative proved difficult to sustain and was ultimately discontinued.

CASEY: Despite these challenges, the group is now well established. Strong group dynamics have developed, which facilitate conflict resolution and collective decision-making. This was particularly important when navigating the complex land-use regulations in Massachusetts, which vary by municipality and require numerous permits related to building and environmental conditions. Having members with relevant expertise was crucial. For instance, zoning regulations did not permit an underground garage, which we accepted. In retrospect, this led to a positive alternative: purchasing two shared electric vehicles for residents' use. Many residents commute to Boston and the wider metropolitan area, often relying on public transportation, which is relatively well developed.

JORDAN: The project was guided throughout by Green Building Standards. We prioritised energy-efficient design, renewable energy systems, and sustainable materials. We believe that environmentally responsible buildings support and encourage more sustainable everyday practices.

CASEY: Returning to the issue of diversity, while the community is relatively homogeneous in socio-economic terms, it does include a mix of ages, including older residents, residents with disabilities, and families with children. In this context, a certain degree of homogeneity has also contributed to stable group dynamics.

AVERY: I agree. We have avoided the kind of sectarianism that can characterise some housing projects. At the same time, participation in a co-housing community like this requires a particular commitment. Residents must be willing to invest in collective processes and engage openly from the outset. Without that, differing values and expectations could undermine the project or reduce it to a purely residential development without the social, sustainable, and ecological qualities we aimed to cultivate.

AUTHOR: It sounds like you have created a distinctive and closely connected community. The shared experiences and challenges you have navigated together clearly play an important role in sustaining these relationships. Thank you very much for speaking with me today.

4.3 School of Design and Sustainability, Berlin, Intentional Community, Bad Zelig, and Intentional Community Großgrünfeld, Germany

4.3.1 Let's play sustainability!

Asmin stands before 12 middle-management leaders from a major German consumer goods company. She is a Professor of Industrial Design with a special focus on experience design and user experience (UX), and she worked as an industrial designer in both the consumer and capital goods sectors at renowned design firms. Since 2020, her work has been mainly dedicated to researching and applying Systemic Design-Oriented Leadership (SDOL) with a focus on sustainability.

SDOL integrates co-created play for eco-social leadership development with methods of systems thinking. In previous UX management workshops, Asmin addressed the challenge that managers do not inherently consider customer-centred, customer-oriented solutions and therefore needed to be made aware of the opportunities of doing so. In her novel ecologically oriented approach, Asmin connects systems thinking and sustainability with ludic interventions through play. She views play as a valuable action that fosters imaginative and active participation in a relatively stress-free situation. She has developed a "serious play kit" that she uses in workshops like this to investigate whether raising awareness and providing training can contribute to a mindset shift and sustainable solutions that transform corporate structures.

The workshop series consists of three consecutive sessions for leaders, each lasting up to four hours. Each workshop begins with a playful, rather simple outside activity – today, a walk in nature near the River Spree. The activity relaxes the managers and triggers awareness of experiences in natural surroundings. The workshop then proceeds indoors, where the participants introduce themselves and receive a brief introduction to systems thinking and systemic design. They were provided with relevant articles on these topics beforehand. The final part of day one comprises a plenary dialogue on the question: *"What does nature give you?"* During the discussion, Asmin notices a shift in the participants' language; the formerly rational and economically focused tone transforms into a more aesthetic and empathetic vocabulary as they describe their experiences in nature.

The second workshop introduces the SDOL play kit. It consists of three differently shaped coloured acrylic elements that can be assembled. Each element represents a specific systemic level: Synthesis, Critique, and Redesign[8]. The elements can be connected, allowing participants to explore the relationships between these levels. The goal of this playful activity is to respond to questions linked to each element, building on the dialogue from the first workshop:

- Orange (Synthesis level): What do you love in nature?
- Yellow (Critique level): What does our society need?
- Blue (Redesign level): What does your organisation or team need for eco-social decision-making?

The play starts with a short introduction to the guidelines. Participants then receive the first question and write their responses on the orange acrylic element. They condense their answers into key terms and can use as many elements as needed. Next, they write their responses to the second question on the yellow element, and finally on the blue element for the third question. When completed, the single acrylic elements are assembled. The participants decide themselves how to connect and relate the individual elements, creating an artifact that visualises the complexity and interconnectedness. Conclusively, they present, discuss, and reflect on their artefact in groups.

On the third workshop day, the participants discuss their responses to the questions in a plenary session. These discussions reveal, among other insights, that systems thinking allows them to recognise similarities between ecosystems and corporate structures, particularly in terms of interconnectedness and self-regulation. The managers primarily view nature from an anthropocentric perspective, associating it with leisure and as a space for activity and relaxation. In addition to fostering discussions, Asmin also observes that the artefacts act as boundary objects[9], reducing the usual competitive behaviour among the managers during the discourse.

Several participants, however, express concerns about the practicality of translating creative ideas into concrete actions. They emphasise the need for clear strategies to bridge the gap between imaginative exercises and actionable outcomes. While the workshop is considered effective in generating visionary ideas and collective insights, the challenge remains in converting these abstract concepts into actionable strategies.

Nonetheless, participants agreed that the "play sustainability" approach is very valuable for generating awareness, fostering self-reflection, stimulating discussion, and motivating them to adopt an ecological perspective. Asmin believes the mindset of society must shift from human-centred to planet-centred sustainability, and the workshop provides an opportunity for societal stakeholders to understand systemic connectedness and explore how emotions and values can be set into action. The workshop exclusively uses recyclable materials and avoids printed materials. Asmin sees this as a small but meaningful contribution toward encouraging leaders to make sustainable changes in their daily (work) lives.

4.3.2 The Aesthetics of Frugality

The year is 2023 BC. Germany is entirely occupied by Capitalism. Well, not entirely… One small village of indomitable dwellers still holds out against the invasion.

This is how a story about this ecovillage could begin. It would tell a tale about a community of 120 committed residents living on the margins of German capitalist society in a small village about 80 kilometres from Berlin. But this is not a story about rage against the “more and more” machine. The ecovillage offers a transformative approach to sustainable design. Gregor, a former designer now responsible for the building circle, tells me how the community started. Forty years ago, he was part of a group of young students and workers searching for alternative ways of living. After years of staying in various locations all over Europe, the group finally purchased an old military site in 1991, making it their permanent home. Their vision was to create a community that meets human needs through simple and sustainable practices and nurtures personal and spiritual growth.

Today, the residents sustain themselves partly through agriculture and by organising seminars. These seminars focus on experimental approaches to cultural design and how they can be integrated into everyday life and contribute to societal change. They often combine hands-on skills with personal growth and community-building. For instance, the “Youth in change” programme trains participants in gardening and sustainable cooking while also educating them in self-development, cultivating sustainable interpersonal relationships, and acquiring facilitation skills to manage collaborative spaces and lead community initiatives. In addition, the ecovillage hosts a variety of cultural events featuring novel artistic expressions and cultural explorations, such as exhibitions, performances, and festivals.

Permanent residents contribute a minimum amount of €563 (approximately USD 655 in 2025) per month, the equivalent of the German unemployment grant, and in return receive free housing and meals. Georg explains that the community operates through sociocracy; decisions are made collectively. A “forum” comprising all residents discusses key issues, guided by advice from different circles: vision and strategy, agriculture, cooking, maintenance, building, and social relationships. Gregor tells me that as the person responsible for the building circle, he gives recommendations for renovations and new buildings based on his experiences.

The residents practise a self-organised approach based on trust, respect, and freedom for everyone. There are no specific rules for individuals, but the community practises empathy, "genuineness", and mutual responsibility. In the communal dining hall, a poster prominently displays the villages guiding principles and the credo: *"We humans live conscious, loving, and free as responsible parts of creation"*. The community kitchen serves free vegan meals to residents, apprentices, and guests, with most ingredients grown locally. The emphasis on mindful living and sustainability is integrated into all daily practices, mirroring the residents' commitment to both personal growth and harmony with nature.

Gregor and I visit the community's common buildings, which are beautifully restored with recycled materials, displaying high creativity and artistic expertise. He tells me that renovating the old structures is expensive and that the requirements to use recycled materials often pose challenges for the building team to develop innovative solutions. The creativity of the building team is particularly evident in the common bathroom, where small tile pieces made from recycled materials create a fascinating pattern. Gregor explains that the intricate design and vibrant colours are the result of countless hours of meticulous voluntary labour from the community members. The design is Gregor's, and he recognises that while the community benefits from an abundance of skilled labour, financial resources can be limited compared to the costs of renovation.

Yet, what is even more remarkable as Gregor discusses the bathroom is the attitude, which reflects another village principle: that beauty does not need to be excessive or luxurious. Instead, the community follows a frugal aesthetics, embracing simplicity, functionality, and resourcefulness. Gregor highlights the intentional use of raw, unpolished materials, reused bricks, recycled wood, and broken tiles, pointing out that the imperfections that come from reuse, wear, and age are welcome, even if he admits he is not entirely happy with the uneven bathroom floor.

Now that I am aware of it, I notice recycled-material art all throughout the village premises. The green spaces are dotted with large sculptures made from metal scraps, plastics, and discarded electronics. In the community buildings, smaller textile and fibre pieces blend traditional techniques with upcycled materials. Many of the artworks display organic forms, and Gregor tells me that these are inspired by artists and designers such as Hundertwasser and Moholy-Nagy[10]. Gregor is especially interested in functional art, for instance furniture from repurposed products, and he also likes artworks that use electronics such as old circuit boards, e-waste, or junk parts to create interactive installations. He thinks that blending art, technology, and sustainability reveals the potential of discarded materials and creates awareness about their usefulness as resources.

What puzzles me, however, is that the communal buildings are much better maintained than the individual houses, which appear somewhat neglected. Gregor explains that the residents cannot choose where they live and how long they stay in a particular house, but that these decisions are made by the forum. Hence, they see no need to upgrade individual houses if they must move out after a period anyway. He is also frank about two other challenges at the village. He tells me, for instance, that for him and some others it can be difficult to be both friends and colleagues within the community: *"the work never stops"*.

Unlike a typical nine-to-five job, leisure time is rare here. In the evenings, other residents often drop by asking for his help, and although Gregor helps them, it can sometimes feel overwhelming. Another challenge is the lengthy application process to become a full resident in the community. The process takes over two years and includes numerous briefings by the forum and applicants are admitted as full residents only when all forum members approve. Certainly, this procedure ensures that the community stays true to its values and principles, but it also means that many applicants drop out before completing the process.

Currently, there is also an age imbalance: the community consists largely of older residents, while younger people tend to stay only briefly before moving elsewhere. The elderly in the village are increasingly struggling with the physically demanding tasks involved in organic farming and eco-construction. Some face health issues and need medical care. Gregor suggests that the village forum might consider new resident structures and become more active in welcoming and keeping younger members and families. Nonetheless, he is satisfied to have spent most of his adult life in this community, where he could use his professional skills and interests to contribute to social transformation. For Gregor, aligning skills, values, daily activities, and sustainable living is key to turning the philosophy of a fulfilling life into a lived reality.

4.3.3 Energy from the Village for the Village

When war broke out between Russia and Ukraine in 2014, Germany's energy landscape was abruptly shaken. Heavily dependent on Russian natural gas – about 55% of its supply – the country faced mounting pressure to diversify and accelerate its *Energiewende* toward renewable energy. While most Germans support the transition in principle, enthusiasm often fades when wind turbines or biogas plants are planned nearby, triggering familiar "not in my backyard" (NIMBY) reactions. Yet the Bavarian village of Großgrünfeld gained international attention as a striking counter-example, embracing a "Please, in My Backyard" (PIMBY) approach by fully powering itself with locally produced solar energy and farmer-generated biogas.

Großgrünfeld is located in northern Bavaria, in the Franconia region, a rural landscape of green hills and farmland, where crop cultivation and animal farming have traditionally been the main sources of income. In 2023, the village had a population of 937. Between 2016 and 2020, international media increasingly recognised Großgrünfeld as a model village for community-driven renewable energy development. The village's initiatives reflect the strong commitment of local stakeholders to renewable energy generation and offer valuable lessons for rural, community-led innovation. This recognition is well deserved, yet Großgrünfeld's transition toward sustainable energy began much earlier, and renewable energy generation has a long tradition in the village. Moreover, the transition encompasses far more than just renewable solutions.

The mayor, Peter Sigmar, tells me how the village gained global attention but more importantly, he explains the community's motto: "from the people, for the people". The village installed its first wind turbine as early as 1923, and already this initiative worked through the collective efforts and investments of the newly established Großgrünfeld Energy Society (GES). Since 2015, four wind turbines produce up to 22,000 megawatt hours of renewable electricity per year. Since 2023, Großgrünfeld meets its energy needs entirely through renewables and produces a multiple surplus of electricity, and the sale of the excess production benefits residents and local businesses. The village has also invested heavily in biogas technology to generate energy from agricultural waste and other organic materials. Local farmers have installed biogas plants; byproducts from the plants are used as fertiliser on the fields, creating a closed-loop system. The CO_2 reduction of the renewables in the village adds up to 955 tonnes per year, and the sale of excess production benefits the community. Approximately 90% of household heating is supplied by waste energy from the biogas plants, saving Großgrünfeld approximately €300,000 annually.

Besides creating opportunities for local development and growth, Mayor Sigmar is very passionate about a special feature of Großgrünfeld's renewable energy production, namely that it is entirely owned by the residents. Sigmar has been mayor in Großgrünfeld for over 30 years and has been re-elected four times. As one of the driving forces of the village's energy reform, he emphasises the power of collective action: *"What is not possible for a single person, many people accomplish together"*. With this quote, he introduces me to Friedrich Wilhelm Raiffeisen, a prominent German social reformer. Even though I first learned about Raiffeisen in middle school many years ago, it is only now that I see his social concept put into practice. Raiffeisen was a pioneer of the credit cooperative movement[11] and founded the first rural cooperative, the "credit union", in 1864. His goal was to combat poverty by helping farmers with affordable loans, laying the groundwork for cooperative banks worldwide. His model, the "Raiffeisen Bank", emphasises solidarity, self-help, and community spirit, and serves today as a blueprint for the citizens' energy cooperative, the Bürgerenergiegenossenschaft Großgrünfeld e.V. The activities of the energy cooperative substantially benefit the village and the surrounding region and contribute significantly to environmental protection and resource conservation. Economic development is increasing, with the village now connected to a motorway, providing easier access to larger cities like Coburg and Bamberg. New projects, such as additional wind turbines strategically placed to maximise energy generation and larger photovoltaic systems including rooftops and open spaces, are planned to further reduce Großgrünfeld's carbon footprint, extend the current energy infrastructure, and support the long-term vision of becoming a model village for renewable energy not only for the region but for the whole country.

Mayor Sigmar is convinced that the energy cooperative's decision to invest in these projects both strengthens their commitment to environmental protection and brings economic benefits to residents and stakeholders.

Culturally, Großgrünfelders are committed to preserving traditional values and keeping local customs alive. Many Franconians have a rather conservative mindset, which is reflected, among other things, in the fact that some residents remain sceptical of the new renewable energy systems. Others are hesitant to join the cooperative because they worry about their investments or question whether profits will be distributed fairly. However, besides being traditionalists, Bavarians are also known for their pragmatism. Many interpret nature as a valuable resource to be protected, and for local farmers, the idea of turning waste into profit aligns well with this practical mindset. Mayor Sigmar assures me that overall, the shift has been well received, with strong support from stakeholders, including villagers, businesses, and local organisations. However, he is aware that while consensus may be relatively easy to achieve in a village like Großgrünfeld, realising renewable energy collaborations in larger German cities where stakeholders are often more antagonistic presents a far greater challenge. Talking to Mayor Sigmar reveals that while intelligently designed technology, careful planning, and thorough organisation are crucial for implementing innovative, eco-efficient initiatives at the community level, shared responsibility and passionate leaders like Mayor Sigmar are equally vital for initiating and maintaining sustainability projects. The public response to the village as an example of effective independent energy management has been extensive, with many other communities adopting its strategies for good reason. Despite challenges, the initiative stands as a model for successful regional change towards a greener future. In doing so, it gives the "German Energy Transition" a face, showing the country's ambitious effort to shift from fossil fuels to renewable energy sources and build a more sustainable society.

4.4 School of Design and Creative Practice, Finland

4.4.1 Sustainable Design Education – A Socratic Quest

The following narrative is based on interviews with professors and students from different design institutions in Finland. It is intended to articulate and explore the dialectical tensions inherent in sustainable design education, as they emerge across pedagogical practices, institutional contexts, and individual perspectives.

Setting: A lively café at the university campus. The atmosphere is dynamic, with students and teachers chatting and having their lunches. Two design professors discuss the school's sustainable design curriculum.

MARTTI: Riku, what's all the fuss about sustainable design? What should the curriculum entail exactly, and is it really working, or just a trend that sounds impressive but delivers little?

RIKU: Well, the curriculum today is based on established sustainable design theories. Integrating sustainable design theories into our curricula encourages students to reflect on problems and develop solutions that address environmental, social, and economic issues. Given the current pressing challenges, teaching both sustainable design concepts and tools is both crucial and timely. Don't you agree this is a good way to reach sustainability?

MARTTI: Yes, I agree theory is relevant, but I have doubts about the theories we are teaching. Both positivistic and overly optimistic concepts take up too much space. We teach, for example, theories that build on the belief that nature's measurability is what counts most, and that smart techno-fixes can move society towards sustainability while supporting current production and consumption patterns. Aren't we just joining the green-washing team?

RIKU: What do you mean? Are you sceptical of sustainable design theories, or are you questioning our curriculum?

MARTTI: Sceptical, hmm, let me think … [*smiles*] In philosophy, a sceptic is someone who questions the reliability of knowledge, and I can find myself somewhat in this description.

Let me make my view clear, so that we are on the same page. Positivist knowledge assumes that statements reflect an objective reality. Our sustainable design curriculum tends to prioritise quantitative assessments of environmental issues, while qualitative aspects come second. While quantitative methods are valuable, many so-called objective results are grounded in assumptions. For instance, mean data, which are not necessarily based on actual conditions, are fed into algorithms in Material Flow Analysis to generate future projections. I also think that important aspects of sustainability are neglected. Let me tell you a story to explain: For years, I cycled through the fields and woods not far from my home in Helsinki. The surroundings were not extraordinary, but it was always pleasant and relaxing to be there, and I often appreciated plants and animals on my rides. A year ago, parts of the land were sold to a paper company, and since then, the landscape has changed radically. Much of the forest has been cleared, and animals and plants have disappeared because their living space is gone. I and many others I talked to feel deprived of the natural experiences we enjoyed earlier.

But let us return to sustainable design theory. Phenomenology, as well as cultural and aesthetic perspectives on nature, is largely overlooked in our sustainable design curriculum. What if we told the students for example to go out and experience the dystopic consequences of environmental degradation, reflect on what the loss of nature does to culture and society, and then develop holistic concepts for radical change? In my view, a paradigm shift is essential to create awareness, responsibility, and engagement among future design professionals.

RIKU: Hmm, you mean like a nature experience in Adorno's negative-dialectics sense? This could be interesting, and sounds all well and reasonable, but I miss design theories for education. What do you suggest as learning material for the students besides their own experiences and empathy?

MARTTI: Juha Varto's ideas are a good starting point toward a critical sustainable design theory. He sees artistic thinking as an onset to develop atunement for surroundings that unfolds as attentiveness, responsiveness, sensitivity, and embodiment. Varto's concept of artistic thinking relies on experience and constantly questions normalised habits. It facilitates developing holistic concepts for radical change.

RIKU: I agree that it might be necessary to supplement our readings. Yet, before the students put complex theories into practice, they first need to understand and reflect on them. How can they develop a holistic perspective without a solid understanding of theories such as systems thinking, lifecycle analysis, and the circular economy? I believe that both academic rigor and instrumentality are essential for a sustainable design curriculum. As a rationalist, I also view intellectual clarity as a fundamental value for successful design. Concepts should be based on logical reasoning and supported by robust evidence. Methods must be applied systematically, ensuring that solutions are both practical and feasible. My teaching objective is to foster a comprehensive and logical approach, equipping students with the skills to develop sound and viable design solutions. Logic does not exclude criticality. Take, for example, Nigel Thrift and his theory on how experiences and practices shape world perceptions and contribute to constructing meaning and knowledge. Thrift[12] shows how embodied and performative aspects of social practices can be dealt with in a rigorous and logical form.

MARTTI: This might be true for this author. However, I am also sceptical about how theory and practice are related. I see programmes full of abstract theories, green buzzwords, and innovative tools. Yet, our cities remain polluted, our resources depleted, and designers are still educated to focus on business rather than sustainability. Where is the impact? Isn't education supposed to do more than just transmit knowledge? If the world remains unchanged, perhaps what's being taught is either ineffective or irrelevant. Are students truly prepared to address real-world challenges after reflecting on theories?

RIKU: I think you are underestimating the role of theoretical education. Yes, transferring theory is challenging and real-life implementation is tough, yet theories generate awareness and the intention to do better, as student responses on our curriculum show. Education cannot guarantee perfect practice, but it sparks critical thinking about the broader sustainability impact of design decisions. Theoretical knowledge can also be an onset for communicating with stakeholders, providing arguments and methods for common planning. In this way, sustainable design theories plant seeds for future practice rather than providing immediate results.

MARTTI: Let's hope that the soil is fertile. Nevertheless, without a value- and practice-based approach, all theory is insufficient. Does this make sense to you?

RIKU: Absolutely. The two are complementary. [Smiles] … As Kant says in the *Critique of Pure Reason* [13]: "Without sensibility no object would be given to us, without understanding no object would be thought. Thoughts without content are empty, intuitions without concepts are blind". For us, this means that without theory, practice is shortsighted or superficial. Without practice, theory remains abstract and disconnected. The challenge is finding the right balance.

MARTTI: Perhaps we should be rethinking design education altogether, integrating values and ethics of sustainability into every design aspect rather than treating it as the specialised niche called "sustainable design"?

RIKU: I wholeheartedly support this, as it aligns perfectly with the transdisciplinary education approach we are advocating at the design department at the school. Sustainability values from all stakeholders concerned, including nature as the third party, should not be treated as add-ons but embedded in the core of design education.

4.5 Intentional Sustainable Community, Oberried, Switzerland

4.5.1 A Double Take on Genius Loci

Every place has its own unique energy, presence, and quality, and serves a different aspect of the soul: development, purification, or simple blessing. First, arrive at Schweibenalp, in your room. Relax, find peace, slow down the pace of life, and settle into your body and soul. When you become curious, the places will begin to call you. Only then should you start your explorations and get to know the different place [14].

With these encouraging words in mind, I arrive at the Center for Contemplative Practice, nestled in the Swiss Alps near Lake Thun. The spiritual community and retreat center is situated 1,100 meters above sea level amid nourishing nature and has flourished for 40 years. The Centre's goal is to foster inner growth, spiritual exploration, and a sense of global community.

The Center for Contemplative Practice was founded in 1982 under the guidance of an Indian teacher and guru. The community promotes holistic living, ecological awareness, and sustainable practices, viewing the Earth as sacred and encouraging a lifestyle that honours the interconnectedness of all beings. The Centre has become a hub for the exchange of spiritual traditions, welcoming teachers and seekers from diverse paths, including Eastern and Western traditions, Indigenous wisdom, and contemporary spirituality.

The premises are well equipped, offering a wide range of seminar rooms and accommodations capable of hosting up to 200 guests during the summer months. The guest house, which provides 100 beds, offers comfortable lodging, complemented by a library and a spacious restaurant serving local organic meals. Accommodation rates are relatively high at approximately USD 160 per night for one person (as of 2023) including three vegetarian meals per day.

However, there are opportunities for discounts: guests can save 10 CHF per day by participating as a "working guest", or they may choose to volunteer. Volunteers typically stay between one and three months and work for six hours a day, five days a week. In exchange for their efforts, they receive free food and accommodation. Working hours might be prolonged during "intensive" periods, and working areas include home care, kitchen, maintenance, and permaculture.

I am unaware of any of these conditions when I take my first tour through the premises, nor do I realise how much work it takes to keep the premises in "natural" shape. However, I am stunned by the natural beauty of the place and the ingenious landscape design. This is embodied "genius loci". Every "Kraftquelle" (place and source of strength) is deliberately enhanced and carefully maintained, displaying mandalas, sacred symbols, or artworks. As I wander through the area, I notice the scent of wildflowers and pine. Nestled in a hollow and surrounded on three sides by tall rushes lies a small pond, its surface still and glass-like. Water lilies drift across the water, their white petals glowing in the soft afternoon light, and I hear the croaks of a small frog or toad. The atmosphere is very peaceful, and I sit on a stone bench for over an hour, contemplating and enjoying. The pond appears sacred, untouched by time. Reflections of the Swiss snow-capped peaks and drifting clouds ripple across its surface. As I continue my walk, I come across more spiritual artworks: stones arranged in spirals, Hindu and Buddhist statues, and carvings etched into the rocks. All of this creates a tranquillity that makes me feel calm and connected to nature.

Leaving the garden, I continue along a quiet path toward a small shop tucked between the trees, its shelves stocked with herbs, oils, and handmade spiritual items. I purchase some essential oils and take in the meditative atmosphere and the friendliness of the people around me. After pausing for a few conversations, I make my way to the dining hall. There, guests gather at long wooden tables; the soft murmur of conversation fills the room, and the aroma of a freshly prepared vegetarian meal drifts through the air.

After the meal, I meet with two managers from the Centre, Anne and Phoebe. As in several other intentional sustainable communities I have visited, they were chosen to speak with me following a steering group discussion and decision. Anne and Phoebe do not live at the Centre but commute daily from the nearby town. They are trained as tourist managers and have a personal interest in spirituality, nature, and human growth, which makes their roles align with both their professional expertise and individual commitment. Both are employed by the "Foundation", the legal administrative entity of the Centre, to oversee individual visits, course activities, volunteers, and other aspects of the Centre's operations. They explain that there is typically a small, dedicated community of residents, along with many temporary visitors, especially during the summer months. The number of permanent residents varies, but the core group generally consists of between 10 and 30 individuals (as of 2023). Residents are actively involved in the daily operations and management of the Centre while also pursuing their individual spiritual paths.

Anne and Phoebe emphasise that everyone is welcome to bring their own religious and spiritual practices to the Centre. Hindu rituals and meditations are performed in the Dhuni (fire tent) both mornings and evenings and are accessible to all. Over the past 40 years, the Dhuni and its surrounding area have become a place of pilgrimage for many visitors, offering spiritual practices for clarity, inner strength, and connection with the universal source.

Anne and Phoebe have not been participating in the rituals recently, as it is July, the peak season, and their workload is high. They describe residents and visitors as generally very nature-conscious and socially responsible. Yet, alongside this emphasis on ecological and spiritual values, the Centre also provides a range of amenities designed to ensure guests' comfort. One example is the small bar where guests can purchase wine and beer. Anne and Phoebe explain that this has been a topic of discussion among residents, as some questioned whether alcohol aligns with the Centre's spiritual values. Ultimately, however, they decided to allow it in order to remain accessible to a wider range of visitors and to accommodate the needs of larger guest groups. In this way, the Centre strives to avoid becoming overly orthodox while still promoting awareness of holistic and spiritual lifestyles.

Anne works in the permaculture section, where eco-conscious living and spiritual practices merge. She appreciates the room for pragmatic choices and hands-on learning, such as developing sustainable food supplies and efficient water and energy use. She tells me she has gained valuable experience in designing efficient garden layouts and creating self-sufficient, resilient environments in harmony with nature. Integrating eco-design principles into infrastructure and daily operations, residents and managers have successfully incorporated local materials and products, solar and wind power, and recycling, composting, and waste-minimisation systems across all activities. Additionally, the Centre offers workshops and educational programmes to raise environmental awareness and encourage sustainable living.

Phoebe highlights the artistic approaches, which use eco-friendly materials and natural forms to aesthetically reflect environmental balance. Art is not only present through sculptures and designs inspired by nature but also through artistic workshops focused on land art, eco-art, painting, sculpting, and other forms, where visitors learn to draw inspiration from the environment.

I am fascinated by the atmosphere of this place and spend the following day exploring its many corners and speaking with residents and visitors. The people who come to the Centre are friendly and relaxed, each engaging with spiritual and sustainable living in their own way. Some seek out practical learning opportunities, attending workshops or joining community activities, while others like me prefer to take strolls on the grounds. Right now, this place feels perfect.

As I leave the Center for Contemplative Practice, I decide to take two small pots of olibanum from the garden shop with me, a small reminder of my time here. The gardener, Rolf, is a volunteer, and unexpectedly another side of the Centre reveals itself. When discussing my "genius loci" impression, he first agrees that the history, geography, and spiritual significance of the place are crucial for creating the unique atmosphere of the place. The alpine setting and the beautiful nature encourage the experience of a serene tranquillity that relaxes and stimulates contemplation.

Rolf then tells me that he arrived two months ago, drawn by stories about this exact character of the place, but so far has had little time to enjoy it as a volunteer. He explains that volunteers play a crucial role in supporting the community and its activities and contribute to all, organising events, maintaining the place, and facilitating the visitors' comfort. Sometimes he works up to eight hours a day, but the extra payment does not make up for the time lost while being here. He also finds that the harmonious "resort" façade hides an economic inequality that most visitors are not aware of. The costs of staying or participating in programmes are relatively high. Those with fewer financial means can only be accommodated by working during their stay, creating a gap between individuals who can fully engage with the surroundings and those who cannot. Accommodation is not equally distributed either: many volunteers live in tents at the nearby campsite, most visitors stay in the main guesthouse, and permanent residents have their own apartments in houses on the premises. Rolf also tells me that his fellow volunteers love the place as much as he does, but that many are disappointed by the mismatch between time for spiritual activities and work overload. The steering group addresses these issues; however, the Centre needs, if not profit, economic progress, and there is no easy way out of this dilemma. At least these concerns are discussed, says Rolf. Yet, he would like to see more transparency on the role of volunteers, for example, not perceiving them merely as service personnel but also as participants in the Centre.

Rolf's arguments prompt me to review the extent to which the concept of *genius loci* and interpretations of the aesthetic experience of nature are affected by affluence and social class. The Centre's unique atmosphere, energy, and character are undeniably shaped by its landscape, yet the design of the grounds also expresses a spirituality imbued with economic privilege. As Smith would argue, through its emphasis on beauty, leisure, and tranquillity, the place does not merely present but actively reproduces a "bourgeois imaginary" of nature, profoundly structured by class-based perceptions (Smith, 1984).

An important ethical consideration is that access to appreciating nature cannot be separated from questions of environmental justice. In 2023, the Center for Contemplative Practice began offering disadvantaged local residents a free stay on the site, and Rolf proposes extending this by granting volunteers an additional week of complementary accommodation. Such measures help mitigate inequalities, promoting more equitable participation. The case of the Centre underscores that access to nature cannot remain an economic privilege of the wealthy but must be claimed and enacted as a collective right.

4.6 Institute for Sustainable Design and Innovation, Intentional Sustainable Community, Singapore

4.6.1 Exploring Five Key Principles of Sustainable Design Education

The following narrative is based on interviews with professors and students from different design institutions in Singapore. It discusses five principles of an integrated approach to sustainable design that links environmental responsibility with economic viability, cultural heritage, and human-centred values, while emphasising real-world engagement, technological innovation, and context-sensitive practice.

At the Institute for Sustainable Design and Innovation (ISDI), sustainability is part of all design, engineering, and innovation activities. ISDI promotes energy-efficient technologies and green infrastructure, facilitating projects in renewable energy, smart cities, sustainable urban planning, and the design of eco-friendly products and services. The college also seeks societal collaboration through partnerships with industry, government, and community stakeholders. The five principles presented in this vignette are based on in-depth interviews with three professors at ISDI and brief conversations with students at the college.

1 Integrating Environment, Economy, and Heritage

Environmental sustainability is at the forefront of urban design. In the past, architectural and historical significance was often seen first, but the economic side is just as important. For instance, in Singapore, the Building Construction Authority (BCA) offers grants up to 600,000 Singaporean dollars (app. 450,000 USD) to help retrofit older buildings to achieve net-zero emissions. This means that for a building or urban block to be preserved, it must become both ecologically and economically viable. The Golden Mile Complex in Kallang is a good example. After it received conservation status from the Urban Redevelopment Authority, it granted additional development rights, such as extra floor space and parking areas, to ensure the project was financially feasible. Conservation in Singapore means not only preserving culture and design in an environmentally sound way but also connecting economic value, heritage, and sustainability, and ISDI education emphasises these values. Students are encouraged to engage with existing buildings instead of starting with empty sites, reflecting the real challenges of contemporary sustainable urban design, including carbon neutrality and environmental resilience.

2 Balancing Human-Centred Design and Nature

The importance of human-centred design (HCD) in architecture and urban planning was already mentioned by Vitruvius (c. 80–70 BC – after c. 15 BC), and by Le Corbusier (1887–1965), who both placed humans at the centre of design. The Indian mandala also reflects human-cosmos relations, for example through the notions of the outer world and inner self, and material desires and spiritual consciousness. So, all design processes start with humans and with a community's needs and desires. Cultural mapping is actively pursued at ISDI, and it comprises a bottom-up rather than a top-down method. This grassroots approach is the foundation for cultural management plans (CMP). Cultural mapping at ISDI is based on the Historic Urban Landscape recommendations established by UNESCO in 2011 (UNESCO, 2011). While originally designed for managing World Heritage Sites, this framework can be applied to various locations. Governments often seek to implement cultural mapping to avoid conflicts arising from top-down approaches, which tend to overlook grassroots input, and ISDI wants to be a facilitator for building trust and dialogue.

Students are encouraged to actively engage with residents and stakeholders, and this collaboration allows them to understand community perspectives before developing their concepts and solutions. Urban sustainable design projects require students to communicate with all relevant stakeholders, and networks with local ASEAN universities enhance learning experiences beyond Singapore. Students are exposed to unfamiliar contexts, such as in Melaka, Malaysia, or sites in Jakarta, Indonesia. After returning to their home institutions, they develop their own studio projects; a recent studio focused, for instance, on rethinking the refurbishment of a wet market in Indonesia. ISDI students who participate in different cultural mapping initiatives collect and disseminate their results on a publicly accessible digital platform. Another successful collaboration was with the People's Park Complex in Chinatown, Singapore, where the Urban Redevelopment Authority and the National Heritage Board informally collaborated with ISDI. The aim of the National Heritage Board is to preserve intangible cultural elements, which cannot be achieved without addressing the tangible aspects as well. They sought student participation to help develop strategies and policies for preserving the area. This project enabled students to work on two scales: one group focused on the broader Chinatown area, while another concentrated on specific buildings.

It is also important to mention that HCD and sustainable design in Singapore are genuinely intertwined with Taoist philosophy. This teaches students to respect the past and 3,000 years of civilisation and to comprehend that design is also a reinterpretation of the past. On the one hand, a scientific understanding of the carbon footprint and other sustainability analyses associated with heritage buildings are crucial, as they inform sustainability decisions and provide incentives for building owners to adopt sustainable retrofitting practices. On the other hand, heritage conservation rooted in Taoist values[15] ensures that physical elements enhance harmony and promote community well-being. Taoism advocates a biocentric view of nature as interdependence between all forms of life. In Singapore, Taoism is part of traditional Chinese beliefs and local culture. Taoist temples like Thian Hock Keng and Wak Hai Cheng Bio, dedicated to Mazu, Goddess of the Sea, embody the city's dependence on maritime environments. Therefore, conservation of heritage in Singapore is not simply about maintaining structures; it is about creating a deliberation between the past and the future. By applying principles of Tao philosophy and leveraging modern technologies like digital twins and BIM, Singapore can continue to honour its rich heritage while promoting sustainable development. In doing so, the city ensures that its cultural identity thrives within a framework that respects both nature and the diverse narratives that define its landscape.

3 Exercising a Real-Life Approach

Architecture and design cannot be done by the book. Theory helps decision-making, but students must be exposed to the messiness of architecture itself. Singapore's architecture regulations are very formal compared to other countries like Indonesia, but to be a successful architect, one has to learn to branch out. Students must widen their areas of interest and skills, expand their understanding, and consider various factors, such as economic pressures and competition, which are prevalent in real life. They also need to understand social and climate factors and find solutions that create a healthy, liveable environment. Only by establishing a suitable environment in this tropical heat can people begin to occupy spaces and engage in sustainable activities. Yet greening in Singapore is not just about creating comfortable temperatures or aesthetically pleasing environments, but also about growing edible food. Providing a reasonable amount of sustainable food in the city is challenging. The sustainable design approach here includes bringing nature to urban environments in ways that make sense for residents. Singapore will not develop extensive biodiversity due to its dense population, but there is room for urban greenery. Residents can engage with sustainability by gardening and growing their own vegetables. When students see iconic places like Marina Bay Sands or Gardens by the Bay, they perceive "greening" in terms of energy-efficient technologies and sustainable urban design. However, they often lack an understanding of how these concepts apply to everyday life, such as rooftop farming or hydroponics in residential buildings. To gain that understanding, they need to immerse themselves in the day-to-day activities of Singaporeans. By collaborating with initiatives like the "Edible Garden City", students can, for example, design community gardens that encourage residents to cultivate plants and vegetables. These collaborations, in turn, foster a deeper understanding of sustainable living at both the grassroots and academic levels.

4 Innovating Technology

While Singapore remains a leader in developing cutting-edge environmental technology, there are still opportunities for improvement. The city is pursuing a 2030 sustainability initiative, which aims to produce 30% of its food, water, and energy domestically. Given its relatively small sise, the country imports water from Malaysia but excels in water management, with approximately 80% of its water coming from recycled sources, including "NEWater" and desalinated water. NEWater purifies treated wastewater through advanced membrane filtration and ultraviolet disinfection. This process produces high-quality clean water, which is used for both industrial purposes and drinking water. For desalination, Singapore uses advanced reverse osmosis technology to convert seawater into fresh water. Since 2023, the city has implemented digital smart water meters equipped with sensors and IoT technology to monitor water usage in real time and alert consumers to leaks or excessive consumption. In terms of wastewater management, Singapore recycles, among other things, greywater through closed-loop systems that treat water from sinks and showers and reuse it for sprinklers, toilet flushing, and irrigation.

While Singapore has succeeded in water management and treatment technology, public discussions about waste management, especially on how to separate waste, are ongoing. Many citizens believe that the current efforts to segregate plastic and non-plastic materials are more symbolic than functional, and in fact, public participation in waste separation remains low. Sustainable designers can play a key role in addressing this challenge, among other things by enhancing public engagement and training, as well as by developing more intuitive and accessible products, services, and systems for solid waste disposal.

5 Educating Responsible Practitioners

Students at ISDI are encouraged to be mindful of the societal and cultural impacts of their designs. They are taught to create solutions that are inclusive, ethical, and responsive to the diverse needs of users. In Singapore, responsibility is deeply intertwined with cultural dynamics and collaborative actions. Students learn to moderate their behaviour, thereby exercising responsibility. Unlike in some Western contexts where individualism, political engagement, and personal expression are often emphasised, freedom is frequently perceived as being intertwined with responsibility in Asian countries. This approach encourages students and professionals alike to avoid conflicts and promote tolerance, especially when working in multicultural settings. The value of collaborative actions is reinforced by real-world experiences such as group work in diverse teams, where individuals from different cultural backgrounds whether Thai, Malay, Indonesian, Chinese, or others work toward common goals.

While a Western approach to ethics is often discourse-dominated, the Buddhist-Confucian perspective emphasises harmony and tolerance. This cultural difference, if not sufficiently acknowledged, can result in friction between Western and Eastern collaboration partners. In Asia, the focus tends to be on conflict avoidance and the adjustment of behaviour, fostering an environment where cooperation is prioritised over confrontation. Furthermore, professional responsibility plays a crucial role in academic and research settings, where individuals are expected to contribute to a collaborative and respectful atmosphere. Educating responsible practitioners at ISDI means linking personal conduct and professional ethics, encouraging students, researchers, and practitioners to work professionally not only for their personal gain but for the collective good. For example, the idea of public exhibitions, where students must present and defend their work, encourages their accountability and responsibility towards society. Moreover, the explicit recognition of societal stakeholders in research reports and public exhibitions highlights the importance of everyone's role in achieving a collective outcome, which reinforces the idea that collaboration and shared responsibility are key to success.

4.6.2 The Tale of Biophilia, Kōng Wèi and Mazu

This narrative is based on site visits to the Queenstown Eco Corridor, which links parks, residential areas, and natural habitats within the Queenstown district, as well as on in situ interviews with residents. Several residents mentioned local traditions and stories related to their surrounding environment, and the narrative is presented as a mythopoetic vignette to reflect these accounts.

Once upon a time, there was a small town by the sea. Its people were fishermen and farmers, hard workers who respected the natural world that sustained them. Yet, as the years passed, business and trade started to flourish, a huge port was established, and settlers from various countries arrived, all transforming the quiet town into a vibrant city. For many years, the future of the city still seemed bright. But growth and neglect of nature went hand in hand. Resources eventually became scarce, and the new enterprises choked the air with emissions. The water turned foul, and waste was piling up in the streets. And still, more people came, drawn by the promise of work and wealth. The city stretched its limits even further, living spaces became more and more crowded, and those who could not afford to build a house had to live in tiny flats, cramped together in the ever-higher growing buildings.

The citizens were modest and quiet people and had long endured the worsening conditions. But one day a group of citizens decided that they could not go on like this. "We cannot forbid anyone to come here," said one, "but our life gets worse, our children ill and our environment polluted. So, what shall we do?" After some back-and-forth discussion, an elderly man took the word: "I know a small island not far from here. They say that there lives a divine healer called Biophilia, who knows the secret to life. Perhaps she can help us". The group decided to send four people on a boat to search for Biophilia and seek her guidance. The party went and indeed, they found the healer. "What you lack is living plants", said Biophilia. "All this concrete will just kill you eventually". "But how can we green the city?" they asked. "Well, I will come with you, stay for a while, and teach you how to do it. Let's make this city a garden!"

With that, they returned to the city. As soon as Biophilia arrived, greenery began to flourish in the streets. She created parks, added shrubbery along the pavements, and planted trees and flowers along the roads. She even introduced greenery on the rooftops and the walls of high buildings. Before long, the air became breathable again, and people began to enjoy taking walks in the city. One of Biophilia's greatest designs was the eco-corridors. She began in one of the city's oldest neighbourhoods, Queenstown, which had deep historical roots in the city and currently a population of over a hundred thousand people, from young families and working professionals to elderly residents. Biophilia first created a ten-kilometre green pathway through the high-rise buildings, linking parks, nature reserves, and green spaces. The eco-corridor included thematic trails, open spaces, and resting zones for residents and visitors. It incorporated native plants, rainwater harvesting systems, and permeable surfaces to improve drainage and reduce heat, thereby enhancing biodiversity, air quality, and the overall environment, and the trails encouraged outdoor activities such as walking, jogging, and cycling.

One day Biophilia designed a scented trail to explore the sense of smell. She planted lavender, jasmine, gardenias, roses, and herbs like mint, coriander, and rosemary, organising the trail into meandering paths that guided visitors close to the plants. Raised beds brought the plants to nose level, making it easier for children and the elderly to appreciate the fragrances. Seating and resting areas allowed visitors to relax and take in the aromatic surroundings.

As Biophilia was almost done, a young woman approached her, holding the hand of a small child. "I just wanted to thank you", the woman began shyly. "My husband and I moved here from Indonesia, we worked so hard, saving every bit to buy an apartment in this neighbourhood. We expected a comfortable home and a good environment for our child, but we never imagined such beautiful surroundings. Sometimes, it feels like I'm back in my village".

While they were talking, the little boy moved around energetically, smelling the flowers and playing in the mild morning sun. Nearby, a group of elderly men, sitting by a small fountain, had overheard the conversation. One of them smiled and remarked, "For us older folks, this place is a wonderful meeting point. We spend our days here when the weather is nice, enjoying the greenery, watching people pass by, and playing Mahjong. It's the best place one could imagine for growing old". Biophilia smiled at their words, content that her efforts had generated meaningful experiences and connections between nature and urban life.

All could have lived happily ever after. But as the city grew and flourished, it soon reached the limits of growth. Amid these adversities, a long-retired and half-forgotten spirit named Kōng Wèi began to stir from his slumber. When Biophilia, the embodiment of nature and harmony, and Kōng Wèi, the personification of spaciousness, first crossed paths, an intense argument between them broke out. Kōng Wèi found that the city's lush greenery would encroach upon the people's living space and that more towering buildings and bustling malls were needed to accommodate the ever-growing population. "Parks are but wasted land!" he insisted. "Parks are the very heart of our city, breathing life into our existence!" retorted Biophilia passionately, and in the months that followed, a spirited rivalry ignited between them. On one side, magnificent skyscrapers and vibrant shopping malls rose high into the sky, while on the other, self-sustaining gardens and serene green spaces blossomed. As time went on, the citizens grew weary of the endless strife. Peace-loving by nature, they felt the weight of Kōng Wèi and Biophilia's disagreements hanging over them like a storm cloud. The desire for harmony grew stronger, for they longed for a place where both nature and community could thrive together.

Seeking help, they went to the city's main spiritual advisor, Mazu, goddess of the sea, and asked her what to do. "You have served me well over the years", Mazu replied, "so now I will assist you". With a wave of her hand, she bestowed upon them the knowledge to create a floating neighbourhood where residential buildings, parks, and malls would rest upon the gentle waves of one of her bays. Once the plan was set in motion, within a few years the new neighbourhood stood steadfast upon the water, with both Biophilia and Kōng Wèi eagerly contributing to the development. However, all citizens were aware that this was a temporary solution, entirely dependent on Mazu's grace. They expressed their gratitude to the goddess for her help and vowed to be mindful of all gifts provided by nature, and to honour its spirits as they always had.

4.7 School of Design and Environment, Shah Alam, Ayer Hitram Forest Reserve, Kuala Lumpur, Malaysia

4.7.1 Eco-Innovation in Action: Experiential Learning Meets Sustainable Design

The following narratives are based on interviews with design educators and experts in eco-tourism from University Putra Malaysia (UPM), a visit with UPM members and AHFR rangers to the Ayer Hitam Forest Reserve, and a discussion with members of the Temuan community in Puchong.

INTERVIEWER: Hello, Selma! It is a pleasure to meet you today. Could you please share a bit about yourself and the work you do?

SELMA: I am teaching product design, furniture design, and automotive design, with each area framed around the core principles of sustainability. At our Product Design Department, every student project is developed with the 17 Sustainable Development Goals (SDGs) in mind, and each project must align directly with at least one of the 17 SDGs. This commitment ensures that sustainability is woven into the fabric of our students' learning experience, and it pushes them to reflect on how their designs can contribute to more sustainable living.

In my work, I frequently collaborate with universities such as Queensland University of Technology in Australia and University Putra in Malaysia. But my work extends beyond academia to include non-academic stakeholders, like businesses and local authorities, who bring real-world perspectives and needs into our projects.

One of the most memorable collaborations we had was in 2018 when we worked with the Langkawi Development Authority (LADA) on Langkawi Island, Malaysia. The goal of this project was to explore sustainable tourism solutions, with our master's students specifically tasked to develop designs that would enhance biodiversity in Langkawi's Geo Forest Park. Over the course of a year, about 30 design students worked closely with LADA, learning about the complex relationship between tourism management and environmental conservation.

Our students were first required to assess various aspects of mass tourism and its impact on biodiversity before proposing design solutions. One key area they focused on was the island's mangrove forests, which function as carbon sinks and play a significant role in reducing greenhouse gases. These forests are also rich ecosystems, home to diverse species of birds, fish, reptiles, and insects. However, tourism activities, particularly boat tours, pose a threat to these vulnerable environments. The sightseeing with boats impacts the mangroves, and the students had to develop concepts that meet these challenges.

Or take the sea eagle feeding as another example. We wanted to explore new ways of feeding the eagles. Feeding the eagles with chicken skin so tourists can watch contributes to them not hunting anymore, and snakes, mice, and other rodents will start to multiply. We first studied the birds' behaviour to understand how an eagle normally hunts. We found that a special type of small fish is part of their diet, and referring to this, one student team designed a floating platform that resembled coral reefs and released small fish at intervals, allowing the eagles in the area to hunt naturally. This eliminated the need for artificial feeding, which can disrupt the birds' natural behaviour. Another team designed an eagle-watching platform that looked like a tree, allowing boats to dock and visitors to observe the birds from a distance without disturbing them.

These projects provided the students with invaluable, first-hand experience and a chance to apply sustainable design principles to real-life challenges. Each group worked within the constraints of the natural environment, considering not only the aesthetics and usability of their designs but also their environmental impact.

We had several groups, and another design example is the monkey food bank. In some parks, monkeys are a huge problem; they like to steal people's bags, snatch food, and even attack sometimes. They know that tourists bring lots of food, and they like to get hold of it. But the food is not nutritious for them, their usual diet being fruits, roots, and leaves rather than cookies and soda. One student group hence developed the food bank, applying biomimicry principles to design an artificial mangrove feeding spot resembling the monkeys' habitat. LADA rangers replenish the food bank with nutritious food suited to the monkeys' diet, discouraging them from seeking out the less healthy snacks tourists often bring. This intervention benefits both the monkeys, who maintain a healthy diet and are not interested in stealing food anymore, and the tourists, who can now enjoy watching the animals' natural behaviours without being harassed for food.

INTERVIEWER: The students' concepts sound really innovative and practical. Do you believe LADA will implement these ideas? Are they more for inspiration, or do they hold potential for actual prototyping and production? And what about the local stakeholders, for example the tourist boat drivers? Would they support the concept of bird watching?

SELMA: Turning these concepts into tangible products often depends on funding, which can be challenging. Moving from concept to prototype requires substantial investment, and LADA, like many organisations, will naturally opt for the most cost-effective solutions. While they may not implement every design, the ideas serve as valuable inspiration, and if they decide to implement a solution, they will certainly align with the local community. For us, it is most important to give our students real-world experience tackling pressing issues. For us educators, the primary objective is not just to spark ideas but to challenge students to think critically about sustainability and economic feasibility. They need to develop solutions that are environmentally responsible but also realistic from an economic standpoint.

INTERVIEWER: In your opinion, what is most valuable for the students to gain doing these kinds of projects?

SELMA: In my perspective, the greatest value is the expansion of their knowledge on biodiversity and its intersection with the tourism industry. Through these projects, they gain first-hand insights into how delicate these ecosystems are and the responsibility we bear in protecting them. For most students, it is a transformative experience. They come in with simplistic views, like assuming monkeys only eat bananas, and leave with a much richer understanding of animal behaviour and ecological balance. These projects are eye-opening, helping students to recognise their environmental responsibility both as future professionals and as individuals. We also included twenty children from local schools in Langkawi in a co-creation session, where they brainstormed ideas for preserving the limestone forests. Our students shared their research with the children, explaining their concepts in accessible terms. This not only allowed our students to reflect on their work but also fostered their understanding of serving the community.

Another important benefit is collaborating with students from other disciplines and backgrounds. Queensland University, for instance, contributes a scientific perspective, while University Putra specialises in ecotourism. Our design students collaborated with biologists and ecotourism students, exchanging insights that revealed the complexity of their work.

One memorable example was when a group of business students from Putra suggested a glass-bottomed boat for observing underwater life in the mangroves. The Queensland students, however, pointed out that the water in these areas is murky, rendering such a feature ineffective. Interdisciplinary collaboration helped our students see the importance of practicality and the potential environmental impact of their designs. The common projects also cultivate their communication skills, as the students learn to convey their ideas effectively to diverse audiences, including non-designers and specialists in other fields. They come to understand how design intersects with sustainable living, benefiting both human and non-human life. I firmly believe this kind of awareness should be nurtured from a young age, ideally well before university. To support this, we proposed the "Young Ambassador" programme to promote Langkawi's Geo Forest Park, involving local children in efforts to raise awareness about conservation. The children produced fantastic ideas, even creating a TikTok video to educate the public on how to responsibly visit the park. This initiative has expanded to include other Malaysian universities and government departments, supported by grants from UNESCO.

INTERVIEWER: To conclude, what motivates your personal engagement to connect design and sustainability?

SELMA: For me, promoting sustainability is both a professional and personal mission. As a designer and educator, I am passionate about encouraging the next generation to prioritise sustainable design and contribute meaningfully to the SDGs. I see this as an opportunity and a duty to inspire my students to think beyond the design itself and consider the broader implications of their work. In doing so, we are not only educating future designers but also fostering a culture of environmental stewardship that I hope will continue to grow.

INTERVIEWER: Thank you, Selma, for such an engaging and insightful discussion.

4.7.2 A living classroom for sustainability education

The Ayer Hitam Forest Reserve (AHFR), located in Puchong, Selangor, is a vital component of the Greater Kuala Lumpur region and one of the last remaining lowland natural ecosystems in this highly urbanised state. Serving as a crucial green lung, AHFR is characterised by rich biodiversity, including various wildlife species, many of which are protected. The forest's unique ecology encompasses flora and fauna from southern, northern, and mountainous regions, highlighting its multifaceted role in supporting biodiversity, culture, and education.

Over the past few decades, significant portions of the forest have been allocated to housing developments, reducing the reserve from its original

4,270 hectares to approximately 1,200 hectares. Strategically located amidst urban development in the Klang Valley, AHFR is interconnected with major highways, emphasising both its accessibility and vulnerability.

Recognising the high conservation value of AHFR, the Selangor state government renewed its lease to the University of Putra Malaysia (UPM) for an additional 80 years in 1996. UPM plays a crucial role in facilitating public access to the forest, focusing on education and conservation efforts. AHFR serves not only as an essential educational resource for the university but also provides tangible and intangible benefits to the community, including clean air, sustainable water sources, and a natural environment conducive to study and recreation. Under the National Forest Act of 1984, AHFR is classified as a research forest, underscoring its significance for scientific inquiry. Research initiatives focus on biology, ecology, environmental studies, and wildlife conservation, with a growing emphasis on ecotourism[16]. The forest attracts both UPM staff and students for scientific research, as well as international visitors from countries such as Japan, Australia, and various European nations.

The educational role of AHFR is particularly impactful for students engaged in practical and industrial training. It serves as a venue for extension activities designed to foster environmental appreciation, enhance observational skills, and promote scientific documentation among youth. Educational programmes for schoolchildren include forestry camps, nature walks, environmental education initiatives, and nature-based activities like team-building exercises and guided visits.

A critical aspect of education at UPM involves teaching students about the ethics and responsibilities of researchers in ecotourism management. The Faculty of Forestry and Environment is dedicated to fostering a deep appreciation for the forest and its ecosystems. In the Department of Recreation and Ecotourism, eco-friendly recreation activities that promote wellness and health are prioritised, ensuring that tourism practices emphasise environmental conservation. This commitment to preserving forests is essential in combating global warming.

In their second year, bachelor's students participate in outdoor recreation activities closely connected to nature, such as camping in the forest. These experiences teach them to guide visitors in responsible practices, highlighting the importance of environmental stewardship. By graduation, students have developed a strong understanding of environmental care. Activities include monitoring trees tied with ropes to prevent damage; every three months, students assess whether the ropes affect the tree's bark, adjusting practices as needed. This proactive approach demonstrates their commitment to being stewards of nature. The concept of stewardship resonates deeply at UPM, reflecting the belief that humans are caretakers of the earth. As nature cannot advocate for itself, it is the responsibility of individuals to advocate for its well-being. Consequently, students have a duty not only to learn about the environment but also to actively care for it.

UPM's outreach extends beyond degree programme students to include primary and secondary school students, individuals from other universities without a forestry background, and local communities, including indigenous populations. Free training sessions are provided to these diverse groups, fostering a more informed and responsible society that values and protects the environment. To cater to different age groups, UPM has developed a series of educational modules. For kindergarten students, the focus is on nurturing a love for nature, especially forests. Engaging activities such as recognising animal sounds, help children connect with the forest environment. For primary- and secondary school students, co-design teaching methods are employed, such as using Lego bricks to create park models, with follow-up activities incorporating natural materials like leaves to be included in the models. At the university level, a multidisciplinary approach integrates expertise from social sciences, arts, biology, and chemistry, broadening students' understanding of ecological principles and fostering collaboration across fields.

Hands-on learning is integral to all UPM programmes. By directly experiencing the environment, students deepen their understanding of ecological concepts. Currently, UPM does not produce commercial products from the forest but focuses on providing hospitality services for outdoor recreation and meaningful experiences. Students often participate in activities like cooking and serving meals outdoors, developing practical skills while emphasising the importance of environmental education.

To protect the environment, visitor numbers to the AHFR are deliberately kept low, with only a limited number of participants allowed on excursions and at events. Visitor feedback varies; some embrace the rustic experience of life in the forest, while others long for more comfort and modern conveniences. This contrast underscores the ongoing challenge of balancing sustainability with diverse expectations. For students, the transition can be tough and many feel out of place in the forest during their first and second year. But by the third year, most have adjusted and feel more at home in the natural surroundings. That gradual adjustment is actually a really important part of the whole experience. And it's interesting that people who are a bit uneasy with nature at first usually start to relax and feel more comfortable after just a few days.

It is really important for all stakeholders involved in ecotourism to understand that comfort often comes with an environmental cost. Resources required for comforts like hot water must be generated, and waste must be disposed of responsibly. In Malaysia, ecotourism is increasingly commercialised, with many locations branding themselves as ecotourism destinations but risking the loss of their natural appeal due to over-commercialisation. Addressing the tension between mass tourism and ecological conservation, Corporate Social Responsibility (CSR) and Environmental Social Responsibility (ESR) are critical concepts in UPM's efforts to promote sustainability among societal stakeholders and businesses. Courses on responsible ecotourism and ESR train operators to avoid commodifying nature to the point of losing its essence. UPM also scientifically measures the limits of tourist access to certain areas to prevent biodiversity loss. Researchers, educators, and planners see it as their responsibility to teach students to think and design for longevity and sustainability. By fostering a deeper understanding of the interconnectedness of environmental issues with human life and well-being, UPM encourages current and future generations to appreciate and protect the natural world while enjoying the benefits of recreational quality of life.

Culturally, AHFR holds historical significance as the habitat of the indigenous Temuan, an Orang Asli[17] community that resided in the area for approximately 400 years. The forest continues to provide essential resources for the community, including food, medicine, and materials. Although the Temuan have recently been relocated to a settlement adjacent to the reserve, they continue to rely on the forest's resources for food, medicinal plants, handicrafts, and spiritual practices. Their deep connection to the forest has fostered a rich material and spiritual heritage, intricately woven into their identity and way of life. UPM students explore the forest to discover plants used in traditional medicine by Orang Asli communities, allowing them to appreciate the practical applications of local flora, such as medicinal plants. This intersection of modern scientific inquiry and traditional ecological wisdom of the Orang Asli creates a unique educational experience. The Temuan see the forest as a living entity with which they share a reciprocal relationship. One of the most striking aspects of Orang Asli culture is their use of resources within the forest. They harvest plants for food, medicine, and tools, always mindful of the ecological balance. When participating in guided foraging trips led by Orang Asli community members, students learn to identify edible plants, medicinal herbs, and materials suitable for crafting tools. These experiences emphasise the importance of biodiversity, showing how each species plays a role in the ecosystem and how interconnected the web of life truly is. Students begin to understand that each tree, plant, and animal has a purpose, fostering a sense of awe and responsibility towards the environment. Moreover, the Orang Asli's approach to land management is rooted in a deep understanding of ecological principles. Practices such as selective harvesting and rotational foraging are designed to minimise impact and allow ecosystems to regenerate. This contrasts sharply with conventional methods of agriculture and resource extraction, which often prioritise short-term gains over long-term sustainability.

Students are immersed in the living knowledge of the Orang Asli, who reveal the forest as a source of identity, skills, memory, and myths, encouraging them to rethink their perceptions of environmental management. They also come to appreciate that sustainability is not merely a set of guidelines and strategies, but a way of life embedded in cultural identity. In this way, education extends way beyond textbooks and field studies.

We meet Rubia from the Puchong Temuan community in her home, and she tells her story: "When I was born, the area was completely covered by a vast forest. The forest stretched as far as the eye could see, and it remained that way for many years. By the time I was around 9 or 10 years old, it was still all forest. All places where we played, went fishing, and searched for fruits and herbs were in that jungle. It was a way of life, and the forest provided everything we needed. We did not have to buy much from outside because everything was available right there in the natural surroundings. From the highway to the hill, it was all dense forest. There were numerous rivers and canals running through the area, and they were pouring with fish. We would often fish in these rivers, catching all sorts of local fish. The whole region was interconnected with waterways, making it a rich source of food. We also hunted boars, various types of animals, and many birds such as hornbills. The forest was so vast that it seemed to go on forever, stretching all the way to the University of Putra Malaysia campus. Going further, there were hills that led deeper into the jungle and my grandfather, who was psychic, once told me that he saw a ghost when he stayed out in the forest overnight, and that he often felt the presence of spirits. He was never frightened, saying that the spirits did not disturb anyone. My grandfather also told us stories of other worlds hidden within the forest, realms that not everyone could see. He said that only certain people had the ability to see these hidden worlds, and that the orang bunian[18], the beings who inhabited them, were just like us. He would then speak of a castle deep in the forest, a majestic place where these special beings resided. It was a world within a world, a place beyond the reach of most. My grandfather claimed he could see it, and sometimes he would disappear, staying in the forest for up to four days at a time. When we would go looking for him, we would find him sitting high up in the trees, as if he had crossed over into that other world, a world where time seemed to work differently. I am not sure whether what my grandfather said was true or simply the product of an old man's imagination, but those stories lived on in our hearts, passed down through generations. Even my daughter, Aisha, heard these tales. To us, they were myths, stories that seemed too magical to be real, yet too vivid to ignore. Whether it was the ghost, the castle in the forest, or the hidden world in the mountains, these were the things that shaped our family's legends. They lived on in the words my grandfather shared, like a fairy tale woven through time".

Rubia's story illustrates how development and urbanisation have reshaped natural landscapes, while also disclosing her ancestors' deep connection to the land. Her narrative evokes a sense of belonging and responsibility that resonates strongly with the students.

At UPM, students explore the cultural significance of various plants and animals in Orang Asli traditions through workshops and discussions. Indigenous narratives enrich students' understanding of biodiversity by placing it within a cultural framework, emphasising the need to conserve not only the species themselves but also the surrounding knowledge, practices, and traditions. This is particularly urgent as many indigenous cultures are at risk of disappearing.

The collaborative projects between Putra University and the Orang Asli also offer students opportunities to engage in research that directly benefits the communities and the ecosystems. These projects often focus on conservation efforts, such as reforestation, habitat restoration, and the establishment of biodiversity monitoring programmes. Students participate in every step of the process, from planning to implementation, working alongside Orang Asli guides who share their insights and knowledge. This collaboration fosters mutual respect and understanding, allowing students to see the value of integrating traditional knowledge with scientific methodologies. As they work together, students learn the importance of adaptive management, a principle that emphasises flexibility and responsiveness to changing conditions. The Orang Asli's deep-rooted understanding of seasonal changes, animal behaviours, and plant life cycles provides students with valuable lessons in ecological resilience.

In this way, the Ayer Hitam Forest Reserve really becomes more than just a research site, but a space where respect for nature is fostered and felt. The forest provides a platform for collaborative learning, where students and communities come together around shared goals and solutions. For many at Putra University, the Ayer Hitam Forest Reserve is a living classroom that offers a rare chance to engage with the interconnectedness of nature and culture. The experiences here go beyond technical or scientific skills they inspire students to think differently about sustainability, to value indigenous knowledge, and to carry those values into their professional work. As one student put it, "The forest makes me feel connected with nature, and by protecting it, I contribute to build a country that is more sustainable, inclusive, respectful, and true to our shared heritage".

4.8 School of Architecture, Kathmandu, Intentional Sustainable Community, Kirtipur, Nepal

4.8.1 Sustainable Planning in and for Nepal – Education for Inclusive Development

The School of Architecture is an influential centre for architectural education in Nepal. Established in 1976, it stands as one of the earliest architecture programmes in the country and remains influential in shaping the landscape of Nepalese architecture. The school is part of the Engineering Institute which is renowned for its technical and natural science faculties, and offers undergraduate, postgraduate, and Ph.D. programs focused on both theoretical and practical aspects of architecture. The Bachelor of Architecture programme is the flagship undergraduate course, structured to provide students with a comprehensive education in design, environmental awareness, technology, and the cultural significance of architecture. The curriculum emphasises sustainable design principles, reflecting Nepal's rich architectural heritage, with coursework incorporating traditional Nepalese building techniques alongside modern design practices. This integrated approach aims to create architects who can innovate while respecting the country's cultural and ecological heritage.

At the postgraduate level, the department offers a master's programme in urban planning, which responds to the growing need for urban development experts in Nepal. This programme covers subjects like urban design, land use planning, and sustainable urbanisation, preparing graduates to tackle the challenges of urban growth in developing cities. The MSc in Energy for Sustainable Social Development (MSESSD)[19] was developed in collaboration with the Norwegian University of Science and Technology (NTNU), integrating sustainability as a central tenet in the education provided. It focuses specifically on social sustainability and energy use, and addresses the urgent need for sustainable energy practices, design, and policies, especially as Nepal grapples with rapid urbanisation, fragile ecosystems, and the demand for energy-efficient infrastructure. Through project-based learning, the curriculum enables students to connect theoretical knowledge with in-situ challenges and real-world applications, educating them to innovate responsibly and prepare them for leadership in sustainable design and planning nationally. Students engage in coursework that explores the social, economic, and environmental dimensions of sustainability, aligned with global frameworks like the Sustainable Development Goals (SDGs), the New Urban Agenda (2016)[20], and the Sendai Framework for Disaster Risk Reduction[21].

The department's focus on sustainability prepares graduates to assume influential roles in governmental and planning bodies where their expertise contributes to shaping policies and projects that are both inclusive and resilient. One of the key approaches in the department is emphasising participatory methods and stakeholder engagement. Students are trained to recognise the importance of involving diverse stakeholders, from local communities to governmental agencies, through coursework and practical fieldwork. This firsthand experience is especially valuable given Nepal's socio-cultural diversity and its varied ecological zones, which include mountainous, hilly, and lowland regions. Each region presents unique challenges in terms of cultural practices, language barriers, and differing levels of education among community members, making stakeholder communication essential. For example, in Mustang, a region in the Himalayas, the interaction and communication styles differ significantly from those in Kathmandu or the Terai (South Nepal), necessitating tailored participatory approaches. To further build resilience and adaptability, students do fieldwork in diverse regions across Nepal, each with its own socio-economic and cultural dynamics. In practical terms, this means that while a solution or communication method might be effective in an urban setting like Kathmandu, it may not work in remote regions like Mustang or in the culturally diverse communities of the Terai plains. Engaging with diverse communities enables students to develop a deeper appreciation for the importance of culturally sensitive and context-specific planning. Through practical, on-the-ground experiences, they realise that effective solutions must be co-developed with the communities they aim to serve, ensuring relevance, usability, and long-term sustainability. Moving towards participatory approaches implies ethical considerations. While well-established globally, such principles are often inconsistently applied in the Nepali academic context. Obtaining informed consent, ensuring voluntary participation, and respecting socio-cultural norms are hence emphasised during fieldwork. By foregrounding these values, MSESSD seeks to instil responsible, transparent, and respectful approaches to data collection, interpretation, and dissemination. Ethical awareness not only protects the rights and dignity of community participants but also enhances the credibility and academic integrity of students' work.

The reflective approach architecture and planning encourage students to move beyond purely technical solutions (Singh et al., 2025), recognising that the success of any project ultimately depends on its acceptability and accessibility among the local users and stakeholders. The collaboration with NTNU has also introduced students to key tools and methodologies for engaging stakeholders. This approach instils a deep understanding of gender mainstreaming and social inclusivity in the students, enabling them to carry forward these principles into professional settings. The real-world training has paid off, as many graduates now work in crucial governmental positions, such as the Ministry of Urban Development and the Department of Housing and Urban Development, where they leverage these skills to make informed and inclusive decisions. While technology plays a vital role in design, the department emphasises that technology must serve people's needs and be accessible.

For instance, a sophisticated waste management system was implemented in Pokhara (southwest Nepal), but its success was limited because the design was not intuitive for the local community. The department uses such examples to demonstrate to students that sustainable design must be functional, approachable, and grounded in user experience. Understanding these principles encourages students to prioritise participatory processes over merely creating advanced products, an approach critical to the success of sustainable architecture.

The department recognises the gap that often exists between theoretical knowledge and real-world practice. To bridge this, the curriculum emphasises experiential learning, allowing students to apply theories in real settings and adjust their methodologies based on on-the-ground feedback. Field-based learning aligns students' expectations with the complexities of professional life, especially in the multi-dimensional field of architecture where social, environmental, and technical aspects must converge. This practice-oriented pedagogy has become increasingly essential and reflects a shift seen across academic institutions worldwide. The case-based learning methodology, which integrates theory with practical engagement, is gaining traction and proves particularly valuable in fields like architecture and planning, where dynamic critical thinking skills are essential.

Looking forward, the Department of Architecture anticipates that sustainability will continue to be a cornerstone of architectural education, not just nationally but globally. Given the environmental crisis, sustainability must become a cross-cutting theme across disciplines, including civil engineering, urban planning, and mechanical engineering. This broad integration would help align educational practices with the broader vision of sustainable development, preparing graduates to design for a world where resources are finite and where development must occur within ecological limits. The department is also preparing to incorporate artificial intelligence (AI) into its curriculum, recognising the technology's potential to support sustainable planning and design. AI could enhance simulation capabilities, enabling planners and architects to model urban scenarios with precision and consider numerous sustainability variables. AI-based simulations could also open new doors for participatory planning, where community members could visualise and interact with future designs through accessible platforms like Minecraft, augmented reality, or scenario-based gaming. These tools would allow the public to offer input on proposed changes to their environment, making architecture and urban planning more transparent and inclusive.

The department acknowledges the growing trend of minimalism, particularly in developed countries, as a response to consumerism. By educating students on waste reduction and sustainable consumption, the department helps foster a mindset that values environmental stewardship and efficient resource use. This awareness-building, facilitated through both theoretical classes and hands-on projects, prepares students to think critically about material use and waste generation in their designs. Sustainable design workshops, for instance, encourage students to devise strategies that reduce resource consumption, from sourcing locally produced materials to minimising the carbon footprint of buildings. Although these projects are often hypothetical, students carry these principles into their professional roles to advocate for and implement sustainable practices.

In conclusion, the Department of Architecture has established a strong educational foundation that fosters both technical competence and social responsibility among its students. The department's emphasis on sustainability, ethical stakeholder engagement, and experiential learning equips graduates to contribute meaningfully to Nepal's development landscape. This not only benefits their careers but also ensures that they are agents of positive change, helping to build a future country that is both environmentally sound and socially equitable. In a world where the urgency of sustainable development is more pressing than ever, the department stands as a model of how educational institutions can drive meaningful change through progressive, practice-oriented pedagogy. The ongoing incorporation of AI and digital tools, along with a commitment to ethical and socially aware practices, promises to keep the department at the forefront of architectural education in Nepal and beyond.

4.8.2 Eco-Transformation in the Making

The following narrative is based on multiple visits to the town between 2016 and 2024, various interviews with local stakeholders, and via student projects developed through the school of architecture. Activities included discussions with representatives of traditional community organisations and local women's cooperatives, as well as observations of ongoing initiatives and design interventions. Together, these perspectives reveal a collective effort toward fostering a greener and more sustainable town, in which contemporary design approaches and student-led projects are integrated with long-standing socio-cultural practices.

Kirtipur is a town located on the southwestern hills of Kathmandu Valley, about 5 kilometres from the centre of the city. The town is a hub of Newar[22] culture and one of the oldest settlements in the valley, famous for its medieval architecture and temples. Today Kirtipur is an integral part of the Kathmandu metropolitan area and faces heavy urbanisation pressures, including pollution, traffic congestion, and the loss of green spaces. These challenges have fuelled Kirtipur's search for green solutions. For almost a decade, residents and local stakeholders have been actively collaborating to implement sustainable strategies, transforming the town into an eco-friendly community. Buildings are being constructed with green materials, while solar panels and rainwater harvesting are becoming a more frequent sight in both residential and public areas. Pedestrian- and bike-friendly zones, with dedicated lanes and walkways, encourage sustainable mobility. The community attempts to follow a circular economy model, focusing on waste separation in households and recycling. Organic waste is composted and used as a natural fertiliser for local farming.

A particularly interesting aspect of Kirtipur's eco-transformation process is the creation of green spaces. The town has identified key areas to establish small parks and green belts, offering residents nature amid urban stretches. The green spaces not only improve air quality and aesthetics but also provide spaces to relax, exercise, and spend time with relatives and friends. The following narrative of Kirtipur's eco-transformation is based on several visits to the town and conversations with local stakeholders, including a representative from the Guthi (a traditional Newar community organisation) and a member of the local Ama Samuha (the women's cooperative). Their insights reveal a collective effort toward a greener, more sustainable town, blending modern solutions with traditional socio-cultural practices.

Kirtipur's eco-transformation started in 2015 when the community, local leaders, and organisations like the Centre for Integrated Urban Development (CIUD)[23] and UN Habitat[24] began to address the increasing waste problem. These stakeholders organised cleanup campaigns, educational programmes, and deployed strategies for the restoration of local water resources. Initial efforts to clean and restore specific areas of Kirtipur gradually expanded into a town-wide programme, though the process was neither smooth nor straightforward. Many public spaces, including open areas, parks, and ponds, had deteriorated into dumping grounds, accumulating waste from both residents and visitors. The pond at the centre of town, once a site of cultural importance and community gatherings, had become green with algae, dying fishes drifting on its surface – an overall hideous sight. Trash accumulated in the water and surrounding areas, creating an environmental hazard and damaging the town's image.

Th merging of inadequate municipal solid waste management and monsoon-related wastewater overflow in the Kathmandu Valley significantly increases the risk of water-borne diseases, including cholera and acute gastroenteritis, particularly during peak monsoon rainfall (Khadka et al., 2022). In Kirtipur, residents, particularly those living near the affected spaces, frequently expressed health and well-being concerns as well. Guthi groups and other community members had started some cleaning efforts, but the impact was limited due to the absence of a unified strategy or systematic waste management plan. Eventually, the leader of CIUD, who had studied Kirtipur's potential as an eco-city in her Ph.D. dissertation (Singh, 2018), suggested a formal cleanup campaign to the town's community leaders. The initiative aimed both to improve waste disposal practices and to generate a sense of environmental stewardship among residents. The first cleanup campaign in Kirtipur began with an ambitious 42-week plan aimed to transform the town's public spaces, starting with the pond and its surroundings. For nearly a year, groups of volunteers gathered each week to remove trash, scrub away dirt, and restore the area. Community involvement was key, with schools, local businesses, civic groups, and residents, even the elderly and children, taking part in the effort. The cleanup was not limited to physical labour. With guidance from CIUD and UN-Habitat, local organisers started training the residents in sustainable waste management practices. Workshops were conducted on how to separate waste for recycling, compost organic materials, and reduce the use of single-use plastics. When the pond area and other public spaces were cleared of waste, the community created small parks for recreational use. Acknowledging these efforts, Kirtipur eventually received financial support from the local government, and through public attention from other areas in the valley, the town earned a reputation as an eco-conscious community.

One of the most significant benefits of Kirtipur's cleanup efforts was the positive impact on local water resources. Historically, Kirtipur had relied on natural springs for its water supply, but these had begun to dry up due to pollution and urbanisation. Waste dumped in open spaces and near water sources had prevented rainwater from naturally recharging the springs, leading to a dwindling water supply. However, as the town's cleanup campaign took hold and waste was removed from critical areas, water began to seep back into the springs. The renewed availability of clean water had far-reaching benefits for the community, improving access to safe drinking water and reducing reliance on external sources. Recognising the importance of protecting their water resources, residents established practices to maintain these areas, limiting waste dumping and keeping the spaces around the springs clear. The revitalisation of Kirtipur's water sources was successful and directly improved residents' quality of life.

However, restoration efforts, particularly in waste management, remain ongoing challenges. The town has now designed a waste management system, which includes weekly waste collection for a nominal fee of around 150 rupees (app 1 USD) each month for each household, and designated drop-off points in public spaces. Three waste collection organisations handle different areas of Kirtipur. The local government hired workers to clean and maintain public spaces, providing employment for locals while ensuring that parks, sidewalks, and other communal areas remain comparatively litter-free. CIUD and UN-Habitat continue to provide technical assistance and training. Through regular workshops, residents and government officials learn new techniques for waste reduction, composting, and recycling.

Kirtipur's environmental transformation has brought many benefits, but various socio-structural issues remain. A key issue is the limited budget allocated to maintenance, which hinders long-term sustainability efforts. Additionally, ensuring consistent participation and raising awareness among residents are challenges. While most residents support the town's sustainability initiatives, there is resistance to change in certain areas. Apathy toward waste management remains prevalent in some communities, and social pressure to maintain cleanliness, while helpful, is not always effective. The high turnover of residents in certain neighbourhoods complicates the maintenance of initiatives, as new residents, particularly those renting for short periods, often show little interest in these efforts. Population fluctuations, combined with limited resources, must be addressed to ensure the long-term sustainability of Kirtipur's eco-transformation.

Addressing social issues within Kirtipur's environmental efforts, Sudha, a member of the local Ama Samuha, identifies several key areas, using the cleaning of the public parks as an example. Along with other members of the Ama Samuha, Sudha plays an active role in cleaning and restoring the town's parks, which serve as a means to bring the community together, especially local women. The park cleaning efforts, which are voluntary, take place every Saturday for two hours. A group of around 67–70 members, mostly elderly women, participates in these weekly cleanups. They gather tools like sickles and grass cutters and work together to maintain the park, which had previously fallen into disrepair. After the cleanup, they socialise, drink tea, and discuss their progress. The Ama Samuha have a longstanding history of contributing to the local community, including the sale of spices, which they produce and sell from their homes. While the cleaning efforts are voluntary, the group receives no external sponsorship for the activity, relying solely on their own resources and initiative. Additionally, the local community is now considering transforming the cleaned areas into picnic spots to attract both local and outside visitors. The idea is to generate revenue by renting out the spaces, which would help fund the parks' upkeep and development. However, problems in implementing this plan concern, for example, access to basic infrastructure such as water, toilets, and sewage management. Meanwhile, the land ownership of the parks has also become a point of controversy. Some areas are owned by the Guthis, while others are government- or municipality-owned. This has created a conflict regarding management and development, with the Guthis seeking to manage the space, while the municipality is also seeking control over the land.

Sudha highlights the central role of the Ama Samuha in cleaning and maintaining the parks, but she makes it clear that the long-term success of these public spaces depends on broader community involvement. Local clubs such as the Guthis and the Youth Club should, in her view, take greater responsibility for enforcing rules and managing the spaces collectively. She sees coordination among these groups as essential to sustaining the parks' functionality and value over time.

From a sustainability perspective, Sudha acknowledges that some waste management practices are already in place. For instance, households are engaged in basic segregation of organic and plastic waste. Yet, she finds the current practices insufficient, particularly in managing the additional waste generated by visitors during social gatherings and events such as picnics, and believes that better waste management is needed to deal with this growing problem.

On the institutional side, Sudha acknowledges the support provided by the Centre for Integrated Urban Development (CIUD), which has actively contributed to park cleaning and the provision of resources. Nevertheless, she believes that CIUD's involvement needs to be deepened, especially to address ongoing challenges such as funding constraints, infrastructure gaps, and unresolved land ownership issues. She suggests that CIUD, alongside other environmentally focused organisations, could take on a facilitative role bringing together stakeholders, mobilising resources, and helping to ensure the long-term sustainability of the parks.

Sudha is optimistic about the transformative potential of Kirtipur's environmental initiatives and what they may bring for the future. She sees opportunities not only for ecological improvement but also for socio-economic benefits, particularly for Ama Samuha and other community-based organisations. If adequately supported, the park could become both a space for community engagement and a source of income. However, she is also realistic about the challenges ahead, noting that achieving this vision will require sustained coordination, infrastructure development, and resource mobilisation. Conclusively, the parks in Kirtipur seem to represent more than just an approach to creating sustainable public spaces for all stakeholders involved, but embody the city's ongoing transition toward sustainability. The efforts of community groups and supporting institutions provide a promising beginning. Whether this eco-transformation can be fully realised will depend on continued collaboration and commitment in the years to come.

4.9 Creative Institute, Northern Highlands Collective, Intentional Sustainable community, Moray Region, United Kingdom

4.9.1 A Win for the Sheep, the Land, and the Pocket

The next narratives draw upon interviews conducted with educators and local stakeholders, in conjunction with an extended on-site visit to the area, offering insights into local practices and perspectives.

As one of Scotland's renowned creative education centres, the Institute has in recent years prioritised sustainability within its curriculum, aligning its teaching with global calls for environmental and social responsibility. Among its initiatives are student projects that explore the circular economy, zero-waste practices, and eco-design in line with hyperbioregionalism[25]. Hyperbioregionalism builds upon bioregionalism by emphasising a hyper-localised focus, often at the level of individual communities or neighbourhoods. This approach promotes the use of local resources, the preservation of local traditions, and the adaptation of practices to specific ecosystems. Students are encouraged to engage with local materials and heritage, exploring how regional identity can inform sustainable design practices. Through gaining ecological literacy, they develop an understanding of native ecosystems, flora, and fauna, which informs their design decisions and ensures ecological coherence. Collaboration with local artisans, Indigenous communities, and cultural historians further enriches understanding of sustainable practices, strengthening the connection between design and regional identity. By addressing the needs of local communities, hyperbioregionalism fosters resilience and well-being, particularly in marginalised areas where limited resources meet abundant local knowledge.

A small rural town in northern Scotland provides a characteristic landscape for design projects that integrate the circular economy and hyperbioregionalism with artistic and creative practice. Since 2016, a regional campus affiliated with the institution has been active in the area, serving as a focal point for its engagement with rural communities. The campus hosts a master's programme in Art, Society, and Publics, which encourages collaborative projects between students, faculty, and local actors. Grounded in the principles of hyperbioregionalism, these initiatives frequently involve partnerships with artisans, historians, ecologists, and other community members. The campus works closely with a diverse network of local stakeholders on initiatives that support sustainability and the preservation of cultural heritage. Community-based projects offer residents platforms to disseminate their narratives and lived experiences, fostering a sense of local identity and resilience. Additionally, efforts to advance sustainable agriculture contribute to the promotion of circular economy practices within the region.

The following narrative introduces Hamish Sinclair, a local sheep farmer who offers insight into his collaboration with design students and academic teams on two key initiatives: a sustainable sheep farming project and the community-based project "Forres Stories". Hamish emerges as a humorous and candid figure, drawing on decades of hands-on farming experience. Grounded in practical knowledge, he balances a deep-rooted connection to the land with an openness to innovation. His reflections illustrate how concepts of hyperbioregionalism and design are perceived from within the community.

Wry and frank in tone, Hamish recounts his experiences on his collaboration: "Working with academic teams and design students is a bit like herding sheep and hoping they know where they are heading. I have been in sheep farming for decades, watching the seasons roll through, working in the thick of mud and fleece, and the rhythm gets into your bones. So, when the idea came up to work with students from the art schools on projects like sustainable sheep farming or 'Forres Stories', I was in doubt. But then I reckoned it could be good for the farm, a way to get a new look at things and maybe bring in a bit of local pride to what we do out here. I figured, why not give it a try? And, well, it has been a ride with some perks, and a few things that make you wonder if these students have ever set foot on a farm in their lives.

"One thing I have come to appreciate is the fresh ideas they bring. When you have been farming as long as I have, you get a bit set in your ways, seeing things in terms of what is practical and what keeps the flock healthy. But these design students see the farm in a way I never would. They look at our practices and start asking all kinds of questions, sometimes ones I never thought to ask myself.

"Take the idea of making the farm more sustainable. For me, sustainability is all about keeping the land and animals in balance, making sure I do no overgrazing or stripping the soil of what it needs. But the students came in with ideas about repurposing waste and turning sheep wool we would normally toss aside into insulation materials or even packaging. I never thought of wool that way. But they came in with samples of recycled products and ideas that actually sounded doable. The students also wanted to get drones up in the air to check out the pastures, map the fields, and talked about some app they could create to track soil health. I think they did not quite understand how unpredictable farming could be, and that the soil is changing constantly, but still, seeing the farm and the pastures from above gave me a new way of thinking about the land.

"Then there was the storytelling side of things. 'Forres Stories' was about capturing the character of the place, showing off the people who live and work here. For me, that is a big part of why we are doing this. My family has been on this land for generations, and we are part of a history and tradition that is not just anywhere. It is rewarding to be able to show that off, and these students are pretty good at digging into the details, seeing the grit and pride in what we do. I would talk about the sheep and our sustainable practices, and they sit there, sketchbooks out, asking about the best and worst days on the farm, or the lambing seasons when one is up half the night in the bitter cold. They wanted to hear the stories behind it all, the way we used to shear sheep by hand earlier, or the way the community steps in when you need an extra hand. To be fair, they managed to capture the spirit of the place, and for that, I respect them. Getting people to understand what it takes to work this land, to raise animals, especially when they come from city backgrounds, felt like giving them a glimpse into a world they did not see before.

"Now, practically, the projects were tricky. Farming does not wait, and deadlines on a farm are set by the seasons and the needs of the animals, and there is no bending those. Lambing season? Never mind it is a Sunday, I am out there with the flock, doing what needs doing. So, when these students wanted to plan things around their academic schedules: 'Oh, we will visit in March and again in October', I had to explain that the farm has its own timeline. Another challenge is how different our goals are. I am focused on getting practical results, while the students are often thinking about ideas, creativity, or research. For instance, they wanted to test some eco-friendly materials on the farm. Great in theory, but when it came to the design students bringing in straw-based sheep shelters, I must smile because these will not stand up well against a Scottish winter. We spent the next day rebuilding with the sturdier old timber we have used for years.

"Some ideas sounded good on paper but made no sense out in the field. Like, they wanted to plant a new kind of grass that was supposed to be 'more sustainable'. But after some back and forth, I had to explain that we had tried that type before, that it did not do well in our soil, and the sheep did not much care for it.

"The gap in understanding is a huge challenge. Farming is hard work, which you do not learn from books or studio work. A few students did genuinely try to get stuck in, but others seemed to have this idealised view of what farming is, imagining we are out there in some green pasture, sheep wandering about, everything picturesque. One time, they suggested a sheep 'tracking' project, attaching GPS collars to see where the sheep roam. But when they realised sheep do not always follow patterns, wandering as they please, they seemed genuinely confused. A few of the collars went missing after the sheep rolled around in the mud. I spent a fair bit of time explaining and sometimes I had to reject ideas that would not work for us.

"However, what I appreciate about working with the students and academics is that they make me think twice. They bring energy, creativity, and a way of looking at things that I have not thought of before. They are open to the kinds of stories and struggles we face, and the storytelling side has given me a way to connect with the community and share what we do here with people who have never set foot on a farm. But I am not always sure that they understand what we need. Farming is its own life, with rules set by the land and the animals, not by grant deadlines or creative visions. I do not need design for design's sake; I need practical solutions that help me work with what I have got. And I need people who understand that when lambing season hits, I am not sitting down to brainstorm. All in all, I'd say working with them is a mixed bag. They mean well, and when it works, it is great. But you have to have patience, and you need to be ready to push back on ideas that will not cut it. Working with academics as a farmer, it is best to know what you want out of it and to be prepared to explain the reality of the land. Sustainable farming is a good thing, but not always easy, especially when making a living it is 'a tight squeeze'. I am on board with sustainable practices if there is fair backing, saying it has to be 'a win for the sheep, the land, and the pocket'".

4.9.2 Finding Garavullin

This narrative is entirely fictional yet based on interviews with several residents of the intentional community, a guided tour through the settlement, and a conversation with one of its most prominent members on the intrinsic value of living beings and atunement to nature as central to the community's way of life.

"I grew up in an orchard near Inverness in Scotland among row upon row of other apple trees. My name is Silvester, and my ancestors were native wild apple trees, *Malus sylvestris*, with a long history. In old Celtic tales, we were said to be the key to the gateway into the otherworld. Like all the other trees around me, I was raised to be undemanding and productive. Orchard keepers pruned my branches, sprayed my leaves, and fertilised my roots, ensuring I grew as they planned. Every tree had its place and purpose: to grow in line and produce high-quality apples. But as I matured, I wondered if this life was the only one possible for me. I recalled that there were places where trees like me grew wild, alongside other plants and creatures. In these places, trees were not moulded into efficient products but grew freely, shaped only by the land and the changing seasons. Other trees in the orchard reminisced about these places as well, though many considered them dangerous and feared they would not survive in the untamed world. Yet, I stretched my branches whenever sensing this wilder world and dreamt of being part of it.

"But one day, in the early spring, my life took an unexpected turn. Two humans from a place called Garavullin came to the orchard. One of them, a man named Gilbert Graham, as I would learn later, was an older figure who moved slowly. He rested his hand on each of our trunks as though he could understand our voices. When he reached me, he stopped and placed his gentle hand on my bark. I felt his kindness, and in that instant, I realised that meeting him would change my life. I reached out, trying to make contact, and he noticed, a smile spreading across his face as he turned to his friend. 'This one is coming with me', he said. 'It's ready for an adventure'. She smiled back: 'You have always had a gift for recognising the special ones'.

"Before I knew it, I was carefully lifted from the earth, my roots wrapped in cool, damp burlap. I was shaken yet tickled as they laid me in the back of their truck, bound for a new life. On the journey, I listened as Graham and his friend Phoebe spoke about their home and their philosophy. Both had been part of the community for decades, working with something called atunement and another practice they called permaculture. Atunement, they said, is key to fostering a collective spirit that rises above individual desires, focusing instead on a greater good. By attuning, individuals and groups can make decisions that reflect their highest values. Listening to their discussion, I could not help but smile to myself. To me, these thoughts seemed so evident; a tree understands this without question. After all, none of us could survive without being attuned to one another and to nature. But perhaps, I wondered, humans must constantly rediscover these truths, explaining them anew each time. Graham and Phoebe also said that atunement serves as a tool for community decision-making in Garavullin. Members can make choices that resonate with shared goals of ecological balance and social harmony. For Phoebe, atunement was personally important for attaining a higher state of awareness and a deeper spiritual connection, while for Graham, atunement was more about listening to the natural world and responding to its rhythms, using it as a spiritual guide for his permaculture practice.

"'Permaculture' was a strange word to me. Hearing it for the first time, I felt both excitement and fear. It sounded like the place in my imagination where plants and animals lived together, free from human control, but it was also a bit frightening, as I was used to having someone governing my life. Would I be able to sustain myself in an environment where all other beings thrived without the structure and guidance I was used to? Later, as I learned more about permaculture, I understood that it does not mean doing nothing but rather following ecological principles that nurture and stimulate natural processes. For Graham, practising permaculture meant working with the local environment, its sandy soils and coastal climate. He applied composting, mulching, and designed forest gardens to regenerate the soil and conserve water, creating a space where species like me could thrive. I felt a deep sense of gratitude when I realised how gently humans could collaborate with nature and all living beings. Maybe it is because he sees permaculture not just as a farming method but as building a respectful relationship with the land and other living beings.

"Upon arriving at my designated place at Garavullin, I was struck by the scene before me. Plants of various shapes and colours grew in a dense, unstructured mix, quite different from the orderly rows I was accustomed to. Surrounding me were herbs, vegetables, towering sunflowers, and wildflowers. The soil was not bare but teeming with grasses, mosses, and clover. Butterflies, bees, and ladybirds moved through the air, while birds perched in nearby branches. 'Welcome to Garavullin', Graham said softly as he planted me beside a lavender bush and a patch of chives. 'Here, we are not forcing you to grow a certain way. Find your place, grow strong, and thrive'. His words filled me with both hope and worry. A pear tree nearby sensed my uncertainty. 'Hello, I am Wilhelmina, and this is permaculture', she said gently. 'Here, every creature and plant has a task, and we work together'. She smiled reassuringly. 'I know it seems chaotic at first, but this is actually how we used to live on this planet'. 'Hi, I'm Silvester', I replied shyly, 'I come from an orchard in Inverness, and all of this is new to me, even though I have been dreaming of a life like this for a long time'.

"As the days passed, I began to realise both the beauty but also the beastliness of my new life. No one came to spray insecticides, and sometimes beetles or caterpillars would gnaw on my leaves or burrow into my bark to lay their eggs. It was unpleasant and hindered my ability to develop fruit. Eventually, some ladybirds, birds, or other creatures would help me to get rid of the nuisance and restore the balance. I reckon that insects are part of the system too, but still, they get on my nerves. A good thing is that the chemical fertilisers from the orchard, which had always been so harsh on my roots, are no longer necessary. I am now surrounded by rich, dark soil, abounding with life. Graham occasionally adds compost and straw to my roots, calling it 'mulch'. It cools and nourishes me, helping me retain moisture much longer. Wilhelmina is my best friend now and, living here for over 200 years, she is really knowledgeable about what is going on between humans and us other beings and explains the importance of each: 'The mulch feeds the soil, the soil feeds our roots, and we, in turn, shade and shelter the plants, creatures, and humans around us'.

"After being here for six years now, I find life much richer, but I also have to do a lot on my own. A hot, dry spell some years ago left me wilting, and I almost died. Graham encouraged me to reach deeper with my roots. As I worked to push them further into the earth, I began to grow stronger. I learned from Wilhelmina and the deep-rooted plants nearby, who thrived even when the rains were scarce. Their persistence inspired me, and gradually, I became more self-sufficient, connected to the land in a deeper way. The best part was that eventually, my roots reached theirs, and we could communicate as our ancestors once had in the forests. Seasons turn into years, and years into decades, and with each cycle I realise more clearly that I am part of something far richer than I could have imagined in the orchard. Each plant, every creature, contributes to a living, breathing community. In winter, as plants die back, they are left to decompose, nourishing the earth for the next spring. One day, Graham again rested his hand on my trunk, and I found myself asking, 'Why do you garden this way? It is so different from the orchard, where everything was controlled'. Graham looked out over the garden: 'Permaculture is more than a method, Silvester, it is a way of honouring life itself. When we try to control too much, we disrupt nature's balance. But by letting nature guide us, everything flourishes in harmony. Humans are not masters of this land but part of it'.

"Now, I have become part of the unpredictable beauty of Garavullin myself. Some days, I miss the orchard's neatness, the sense of order and control, but I understand that it is the diversity here that makes us resilient. If one plant struggles or withers, others spring up to fill the gap, and the garden carries on. There is a deep kind of balance in that. In autumn, when my branches are heavy with apples, Graham and his friends come to collect my fruit. They value each apple, all grown on a tree that has flourished in freedom. I rest in winter, and every spring, I grow new branches and leaves out of my body, and the birds return to me. Here, in this ever-changing, vibrant community, I have found my place, and I feel honoured to be part of the permaculture garden at Garavullin, which I now call home".

4.10 Solstrand Institute of Design and Sustainability, Intentional Sustainable Community, Norway

4.10.1 Sustainable Design as Transformative Force

The following narrative is based on various interviews with educators in Norway working in the field of sustainable design, as well as resident interviews and site visits conducted within an intentional community.

Since the early 1990s, the Solstrand Institute, a leading Nordic design school, has embedded sustainability across its research, teaching, and innovation initiatives. Education spans energy-conscious design, participatory urban development, and tools that foster community engagement in cultural and social sustainability. Their educational model promotes interdisciplinary collaboration and critical inquiry, incorporating approaches such as "undesigning" to challenge and rethink unsustainable norms. Through hands-on coursework and external partnerships, students are equipped to address real-world sustainability challenges.

Signe, a senior academic at the Solstrand Institute has seen her research interests evolve significantly over the course of her career in sustainable design education and practice. She explains that her early work focused on sustainable behaviours and practices, particularly how individuals and communities might adopt more environmentally conscious routines in their everyday lives. Over time, her focus shifted toward structural aspects of sustainability exploring how systems and institutions can be designed to support and enable broader transitions toward sustainability.

Her research has examined the relationship between everyday consumption and design, investigating how design can serve as a tool to encourage more sustainable lifestyles. This included exploring the influence of industrial designers on consumption patterns and the types of products and systems introduced into daily life. By addressing both individual and systemic dimensions of consumption, this body of work highlighted the critical role of design in shaping more sustainable futures.

In later stages Signe's research expanded into additional domains within sustainability, including energy use and management in non-residential buildings. This line of inquiry focused on the relational dynamics between key stakeholders such as building managers and users and how these interactions influence patterns of energy consumption and overall efficiency. The research highlighted the complex interplay of organisational structures and behavioural factors that shape energy-related outcomes in institutional settings.

Another focus of interest is citizen engagement and participatory processes in the context of sustainable city planning. A particularly notable project involved collaborating with residents to co-develop future imaginaries for a mid-sised Nordic city. Participants were encouraged to envision possibilities for various urban areas including the city centre through informal, community-driven workshops. To support these engagements, digital tools were created to capture and share citizens' ideas. The collected input was synthesised into a collective vision, which was later presented to local planning authorities as a community-based proposal for future urban strategies. In parallel, architectural teams were invited to produce their own development scenarios, providing a professional perspective on the future of key city areas. The findings from this initiative fed directly into local planning discussions and policy development. Themes emerging from citizen input included a desire for greener urban environments with more trees, parks, and natural elements.

Another initiative focused on incorporating sustainable and ecologically rich design principles into the planning of public spaces. It also investigated how various professional groups perceive and value urban nature. For example, while landscape architects tended to view elements like trees as vital components of urban ecosystems, other stakeholders approached them primarily from technical or infrastructural standpoints. These differing perspectives underscore the importance of fostering integrative approaches in urban design that balance ecological, social, and functional priorities.

Education is a central dimension of Signes's work. Teaching sustainable design, she emphasises, involves balancing critical thinking with the development of practical design competencies. One ongoing pedagogical challenge she identifies is determining the appropriate moment to introduce critical sustainability perspectives whether early in a student's academic journey to shape foundational thinking, or later when learners are more equipped to engage with complex, systemic issues. She also reflects on the variation in learners' preparedness and engagement, particularly in relation to their academic and professional backgrounds. In a course focused on systems-oriented approaches to sustainability, some students demonstrated for example a strong capacity for critical reflection, while others encountered difficulties in grasping theoretical concepts. These differences highlight the need for adaptable teaching strategies that respond to a range of learning needs.

Another key challenge identified in the development of sustainability curricula is the need for meaningful integration across academic programmes. While embedding sustainability as a cross-cutting theme can reinforce its significance, there is a risk that it becomes superficial, treated as a checklist rather than a deeply embedded pedagogical commitment. To address this, sustainability must be positioned as both a transversal and a standalone area of study. Dedicated courses and modules can provide the depth and analytical focus required to engage with complex sustainability challenges, ensuring that students acquire not only exposure to the topic but also the critical tools and conceptual frameworks to address it meaningfully.

Collaboration also plays a central role in Signes's pedagogical and research approach. Interdisciplinary teamwork particularly among students from fields such as design, engineering, and the social sciences, creates opportunities for richer engagement with sustainability issues. Joint projects encourage cross-pollination of methods and perspectives, cultivating an environment of mutual learning and innovation. Furthermore, partnerships with external actors, including companies, public agencies, and community organisations, provide a valuable bridge between theory and practice. These collaborations not only enhance the relevance of academic work but also contribute to locally grounded and socially responsive sustainability solutions.

In considering the role of design as an agent of transformation, Signe highlights the concept of "undesigning". This approach moves beyond the traditional focus on the creation of new products and services, emphasising instead the intentional dismantling, phasing out, or reconfiguration of unsustainable systems, practices, and infrastructures. Undesigning invites a critical rethinking of what should no longer be maintained or reproduced in the context of sustainability challenges. Rather than viewing design solely as a tool for innovation or production, this perspective reframes it as a means for questioning and undoing existing structures that contribute to environmental degradation, social inequity, or systemic inertia.

In both research and pedagogical contexts, the notion of undesigning signals a paradigm shift from incremental improvements to more systemic, transformative approaches. It challenges prevailing assumptions within the design field and encourages practitioners, educators, and students alike to confront the embedded values and power dynamics within design processes. As such, it opens up space for reflexivity, critical engagement, and long-term thinking qualities increasingly recognised as essential for addressing complex and interlinked sustainability issues.

Signe's expertise spans several domains, including energy systems, urban development, and higher education, reflecting a commitment to both theoretical and applied dimensions of sustainable design. Her work draws connections between practice and policy, design and systems thinking, and individual behaviour and institutional change. Broad in thematic scope, her approach remains nevertheless grounded in a normative value set and an ethical stance. As she puts it: "Central to my work is to explore how design can shape sustainable futures. Through research, teaching, and collaboration, I strive to contribute to creating a more sustainable and just world".

4.10.2 Fading Green

In the early 1980s, at the beginning of Norway's ecological movement, a group of forward-thinking individuals gathered in Elvheim, a small village near Oslo. Their ambitions went beyond the environmental, political, or economic concerns of the time: they envisioned a way of life rooted in spirituality, communal living, and a profound connection with nature. This loosely formed collective, an early precursor to what would later become an ecovillage, brought together a diverse group of philosophers, artists, architects, designers, and environmentalists. They found inspiration in a harmonious blend of ancient traditions and emerging ecological knowledge, and were drawn to the deep cultural heritage of the land, particularly the ancient Norse reverence for nature and the land's perceived spiritual power.

The founding principles of the collective were based on the idea that living in harmony with nature was not only a matter of ecological sustainability but also a path to spiritual enlightenment and community existence. Temporary shelters were constructed using locally sourced, sustainable materials such as wood, stone, and grass. While most members maintained their primary residences in Oslo, they frequently returned to the site to participate in spiritual ceremonies that celebrated the changing seasons, solstices, and equinoxes. The heart of the settlement was a stone circle and a communal building that served as a central gathering space for meditations, community meetings, and seasonal celebrations, fostering a shared sense of purpose and connection.

Until the early 2000s, the site remained modest in scale, uninhabited on a permanent basis, and only loosely connected to the Elvheim municipality and its residents. However, between 2002 and 2003, a Foundation was established, eventually comprising over 100 members from Oslo and its surrounding areas. The objective of the Foundation was to formally create an ecovillage and construct approximately 50 homes. The ecovillage was officially founded in 2007, facilitated by the Elvheim municipality's decision to allocate a nearby farm as the foundation for its development. It soon attracted an increasing number of individuals and families searching for an environmentally sound lifestyle.

Early development of the ecovillage focused on energy-efficient housing, organic farming, and community-centred governance structures. Another milestone occurred in 2018 through increased financial investment, which led to the expansion of sustainable housing and infrastructure improvements. Between 2018 and 2024, the ecovillage developed into a model for sustainable development, integrating ecological building techniques, renewable energy solutions, and collaborative decision-making processes. The community adopted advanced infrastructure technologies, including solar panels, wind turbines, and composting toilets, to enhance sustainability and self-sufficiency while reducing dependence on external resources.

Initially, these technology upgrades were seen as an extension of the village's ecological and spiritual values. Solar panels were considered not just a practical tool, but an embodiment of the village's desire to live harmoniously with nature. Similarly, rainwater collection systems and organic farming practices were implemented not only to minimise the village's environmental impact but also as part of a permaculture philosophy, emphasising sustainable and regenerative land use.

Yet, the advancement of technological innovations also gradually shifted the community's dynamics. A more pragmatic, efficiency-driven mindset emerged, leading the community to prioritise infrastructure optimisation. Meanwhile, cultural and spiritual rituals, along with community gatherings once integral to village life, were progressively phased out.

Externally, the village evolved into a model for integrating ecological design within a community framework. Researchers in sustainable urbanism, environmental sociology, and participatory community planning began to study it as a case for practical ecovillage design. It received public and government funding, which allowed for more extensive technological upgrades. New eco-friendly building materials were introduced, and the village's systems became more sophisticated. Vertical farming systems, biogas generators, and smart grids were added to improve energy efficiency and reduce the ecological footprint.

With the increased influx of affluent residents, the community's focus shifted once more. The central stone circle, the heart of collective gatherings in the 1980s, was no longer of spiritual significance. A new structure for hosting visitors of educational sustainability-focused workshops was built. Yet, as the ecovillage gained greater external recognition, conflicts intensified, particularly from 2021 onward, between supporters of mainstream production and consumption practices and defenders of alternative sustainable lifestyles, especially between developers and longstanding community members. These tensions eventually threatened the very existence of the community.

In 2024, the property developers associated with the ecovillage declared bankruptcy, leading to significant uncertainties about its future. But the village endured, and in December 2029, it began a third building phase to construct an additional 130 homes. This development includes sustainable residences designed to minimise environmental impact while offering a high quality of life. Residents are encouraged to participate in cooperative organic farming initiatives, reinforcing a regenerative lifestyle.

As of 2026, the ecovillage stands as a highly functional, self-sustaining community, though one that differs significantly from the original vision conceived nearly five decades earlier. It has evolved into a contemporary technological showcase of what a "green" community can look like, equipped with the latest innovations in renewable energy, zero-waste systems, and sustainable architecture.

According to a resident from the earliest days, the spiritual practices that once shaped daily life have gradually faded, giving way to a more technology-oriented approach to sustainability. What began as a sanctuary for reflection and a deep connection to nature has transformed into a centre of environmental innovation. While many see this evolution as necessary progress in the face of global ecological challenges, the resident reflects on the transition with both appreciation and concern, questioning what is lost when a spiritual connection to nature is replaced by instrumental calculus.

4.11 Concluding Remarks

Chapter 4 concludes by reflecting on how the book's central themes: human–nature relationships, human and professional responsibilities, and the role of products, are expressed across the narratives. These accounts span a diverse range of contexts, including institutions of sustainable design education and intentional sustainable communities. While business, policy, and non-governmental sectors are highly relevant to these discussions, they were not the primary focus here, due to both the scope limitations of the book and a deliberate decision to concentrate on grassroots and educational settings. Undoubtedly, many more perspectives could have been included, and the narratives presented are illustrative rather than exhaustive.

Despite their differences, common tendencies appear across the examples, offering valuable insights into the future development of sustainable design concepts, methodologies, and practices, and their growing connection to environmental ethics.

A common motif in the narratives, especially those concerning intentional sustainable communities, is the change in human–nature relationships from a spiritual connection to the environment towards a more utilitarian and efficiency-driven orientation. This is also described as dialectical tensions encompassing eco-technological advancements. While eco-technological innovations enable the implementation of sustainable lifestyles, they simultaneously risk overshadowing ecological wisdom, traditions, and values.

The desire for atunement with nature, for example, by integrating environmental progress with the spiritual and cultural dimensions of human–nature interactions, was highlighted in some narratives. In this context, "integration" emerges as a central notion for environmentally ethics-inspired sustainable design, which will be examined further in Chapter 5.4.

While nature is frequently interpreted pragmatically, in terms of its utility or as an object of stewardship, some respondents emphasise the significance of non-instrumental nature encounters. The value of these experiences lies in their freedom from intentional utility, fostering aesthetic and contemplative engagement. In turn, such encounters can cultivate new ways of perceiving and relating to non-human beings.

In design curricula, changing human–nature relationships is discussed in terms of both the benefits of eco-technology, and design for sustainability, as well as a critical examination of the structural impacts of these "improvements" such as rebound effects. Concepts like "undesigning" are gaining momentum as strategies to phase out unsustainable practices and develop more viable alternatives. Rather than focusing solely on creating new sustainable products, many design institutions advocate for (radically) rethinking existing production and consumption systems and their structural impacts. The integration of sustainability across design education is seen as crucial, ensuring that students critically assess consumption patterns, eco-design options, and the systemic effects of their work. Complementing this, minimalism and conscious consumption are positioned as necessary design strategies for reducing environmental footprints and supporting a sustainable transition.

In both institutional and community settings, products and services are regarded as integral to everyday life. When designed thoughtfully, they can actively support and enhance sustainable practices. Well-designed sustainable products contribute to environmental strategies by promoting responsible consumption and encouraging both individual and collective shifts toward more sustainable lifestyles. Beyond their functional role, such products also serve as tools for education, awareness, and action.

The narratives illustrate how sustainable design can foster a sense of community and solidarity not only among individuals, but also in their relationship with nature. Several examples reflect a relational understanding of sustainability that connects and mutually reinforces social and ecological well-being. Within this relational framework, values such as responsibility, fairness, well-being, and ecological interconnectedness are central. These values are not interpreted as abstract ideals but are embedded in everyday practices and community interactions, influencing how sustainability is enacted. In this way, sustainable design contributes indirectly to ethical awareness and helps cultivate the social and environmental fabric of which it is a part.

Designers play a role that extends beyond aesthetic and functional considerations. In institutional and intentional sustainable community settings alike, they are recognised as key agents of change, whose effectiveness depends on their ability to engage with users, understand their needs, and facilitate meaningful interactions with sustainable products and services. By embracing these responsibilities, designers can drive systemic change, promote equity and access, and contribute to a more sustainable future. In doing so, they are obliged to balance innovation with ethical and environmental responsibility, actively engaging in interdisciplinary collaborations. Additionally, they are tasked with addressing social and political dilemmas, ensuring that their designs cater to diverse communities while integrating sustainable practices. Many respondents emphasise that design should not only serve individual needs but also contribute to systemic change by fostering equity, access, and environmental stewardship.

The findings from the narratives provide valuable insights from practice, highlighting the importance of connecting sustainable design with environmental values and ethics to foster more ecologically sound ways of living. A critical challenge explored in the following chapter is how sustainable design research, curricula, and practice can integrate ethical values for the environment while translating theoretical concepts into relevant and applicable real-world solutions.

Notes

1 A story refers to the sequence of events that occur, while a narrative encompasses the way those events are structured and presented. This includes elements such as theme, setting, character development, plot construction, and event sequencing (Abbott, 2008). Storytelling communicates a narrative to an audience, often employing literary techniques such as metaphor, tension, climax, and closure.

2 Albert Schweitzer (1875–1965) was a German moral philosopher, theologian, and ethicist, best known for formulating the ethical principle of "Reverence for Life" (*Ehrfurcht vor dem Leben*). His philosophy asserts that all living beings possess intrinsic value through their existence and that humans have the moral obligation to respect, preserve, and care for life, regardless of its utility.

3 Excerpt from the original speech by Windelband (1894): *"Die Stärke der Naturforschung liegt auf der Seite der Abstraktion, diejenige der Geschichte auf der Anschaulichkeit. … So reden zu uns durch den Mund der Geschichte vergangene*

Sprachen und vergangene Völker, aus der Vergessenheit zu neuem Leben erstanden, ihr Glaube und ihre Gestalten, ihr Ringen nach Macht und Freiheit, ihr Dichten und Denken. Wie anders ist die Welt, welche die Naturforschung vor uns aufbaut! So anschaulich ihre Ausgangspunkte sein mögen, ihre Erkenntnisziele sind Theorien … sie lässt das einzelne sinnliche Ding, das entsteht und vergeht, als wesenlosen Schein hinter sich und strebt zur Erkenntnis der gesetzlichen Notwendigkeiten auf, die in zeitloser Unwandelbarkeit über alles Geschehen herrschen … Gleichgültig gegen das Vergängliche wirft sie ihre Anker in das ewig sich selbst Gleiche; nicht das Veränderliche als solches sucht sie, sondern die unveränderliche Form der Veränderung".

4 Nomothetic knowledge is often understood as the same as epistemic knowledge, but epistemic includes both nomothetic (focused on general laws) and idiographic (focused on unique cases) knowledge. The term epistemology covers both general rules and specific, context-based knowledge (Salvatore et al. 2010).

5 In Hinduism, Aghoris are sect of Shivaism, i.e. the worship of Lord Shiva. Aghoris practise unconventional spiritual rituals aimed at transcending societal norms and achieving enlightenment. They believe that everything, even what seems impure, is a manifestation of the divine. Aghoris are ascetics, dwelling in cremation grounds or other places considered impure. Their diet can comprise rotten food, faeces, urine and putrefying human corpses. Aghori practices are often met with fear, disgust and scepticism by society. However, their approach to social transgression makes them a compelling headline of this vignette, challenging ingrained social taboos, norms, and thinking.

6 The fundamental opposition between the pure and the impure permeates the entire Hindu caste system. The Brahman caste (priests) is at the top of the social hierarchy, followed by Kshatriyas (warriors), Vaishyas (originally peasants but later merchants), and Shudras (artisans and labourers). Communities that fall outside the four main categories are deemed ritually unclean and referred to as "Untouchables" (Dalit). Tribal groups and foreigners are considered as casteless. Social stratification is based on the level of "impurity" determined by the cast member's traditional contact with various "pollutants" such as excrements, blood, menstrual flow, saliva, dung, leather, dirt, dogs, pigs, corpses, and contact with lower-caste individuals. Intercaste restrictions are enforced to prevent the relative purity of a particular caste from being tainted by pollutants associated with lower castes. All Hindus are expected to continually strive to protect, maintain, and enhance their personal purity. This purity is compromised by contact with pollutants. Even the smallest particle of the human body contains the substance of the person and particles from lower-caste individuals must be strictly avoided.

7 Workers who clean toilets, drains, sewer systems, and septic tanks often dealing directly with human excreta are frequently referred to as "Bhangi", a derogatory term used for members of the lowest castes of Untouchables in India.

8 "Synthesis" relates to problem framing and divergent thinking, "Critique" to identify the leverage points, and "Redesign" to create the system intervention.

9 Boundary objects are implemented in, for instance, participatory design. The have the potential to connect diverging views by providing a common ground for discussion (Keitsch, 2015)

10 Hundertwasser is renowned for his colourful architectural designs and his rejection of straight lines, embracing organic forms to express ecological values and human individuality. Moholy-Nagy was a central figure in the Bauhaus movement, advocating for the fusion of art, technology, and industry. See also Lucie-Smith (2020).

11 For info on Raiffeisen, and credit cooperatives see: Credit union | Savings, Loans & Membership Benefits | Britannica Money. Retrieved 20 May 2024, and Anheiner H. & Toepler, S. (2010) *International Encyclopaedia of Civil Society*, Springer, online.

12 Thrift, N. (2007). Non-representational theory: space, politics, affect (1st ed.). Routledge.

13 Kant, I. (1998). *Critique of pure reason*. Cambridge edition of the works of Immanuel Kant (A51/B 75).

14 Translated from community guide: "Jeder Platz hat seine ihm eigene Energie, Präsenz und Eigenschaft und dient einem anderen Aspekt der Seele: Entwicklung, Reinigung oder zum einfachen Segen. Komme zuerst einmal auf der Alp, in Deinem Zimmer, an. Entspanne Dich, komme in die Ruhe, verlangsame das Tempo des Lebens und finde Dich in Deinem Körper und Deiner Seele ein. Wenn Du neugierig wirst, werden die Plätze beginnen Dich zu rufen. Dann erst beginne mit Deinen Erkundungen und dem Kennenlernen der verschiedenen Orte". (No reference due to anonymity).

15 The main Taoist values are harmony with nature, living in balance with the natural world and respecting all forms of life; simplicity, reducing desires and material pursuits to achieve inner peace and contentment; *Wu Wei* (non-action), flowing with the natural order of the universe rather than trying to control or force outcomes; and compassion, practising kindness, empathy, and mutual support to promote harmonious coexistence (Cheng, 2024)

16 "Ecotourism" refers to travel to exotic and often endangered natural environments. When simultaneously promoting conservation efforts it is described as seen as a form of ecodevelopment, offering a practical and effective way to achieve social and economic progress (Ceballos-Lascurain, 1996).

17 The term "Orang Asli" in Malaysia refers to "indigenous people" and encompasses a diverse array of groups and tribes, each with its own distinct language (Endicott, 2016). These communities have inhabited the forests, coastal regions, and jungles of Malaysia for over 60,000 years, possessing eco-wisdom that is deeply rooted in their connection to the land. Their reverence for the natural world manifests in rituals, stories, and daily practices that honour the interconnectedness of all living beings. Orang Asli communities traditionally engage in sustainable agriculture, hunting, and gathering, ensuring the conservation of natural resources for future generations. They regard themselves as stewards of the environment, recognising that every action has consequences for ecological balance.

18 Orang Bunian (Malaysia, Brunei, and Indonesia) translates to "hidden people" or "whistling people", sometimes compared to elves. Orang Bunian are supernatural beings often depicted wearing traditional Southeast Asian dresses. They resemble humans in appearance. https://vocal.media/horror/whispers-of-the-forest-the-enigmatic-orang-bunian-of-malaysia. Retrieved 20 November 2024

19 The author of this book was the Norwegian project leader of MSESSD, so the programme is not anonymised.

20 The New Urban Agenda, adopted at the United Nations Habitat III Conference in 2016, and later endorsed by the UN General Assembly, outlines a global vision for sustainable urbanisation. This agenda emphasises the transformative potential of well-planned and well-managed urbanisation to drive sustainable development. It promotes inclusive, resilient, and safe cities by integrating environmental sustainability, social equity, and economic growth, aiming to make cities centres of opportunity and well-being for all.

21 The Sendai Framework for Disaster Risk Reduction (2015–2030) addresses the three dimensions of disaster risk in order to prevent the creation of new risk, reduce existing risk and increase resilience. https://www.undrr.org/publication/sendai-framework-disaster-risk-reduction-2015-2030. Retrieved 20 November 2024.

22 The Newars are the indigenous inhabitants of the Kathmandu Valley. Their language, Newari, has Tibeto-Burman origins, and their beliefs blend Hindu and Buddhist traditions. The Newars are renowned for their skills as craftsmen and farmers. Major festivals, such as Indra Jatra and Buddha Jayanti, feature colourful processions and traditional dances. Newar architecture is famous for its intricate

wood carvings. Newar culture emphasises a strong community spirit, deep-rooted social ties, and communal celebrations (Gellner, 1986).

23 The Centre for Integrated Urban Development (CIUD) is Nepalese non-governmental organisation to foster sustainable urban development through innovative, community-centred approaches. CIUD projects focus on waste management systems, green spaces, and eco-friendly infrastructure in urban and peri-urban areas. CUID is partner with UN Habitat. https://ciud.org.np/. Retrieved 20 November 2024.

24 UN-Habitat is a United Nations agency focused on sustainable urban development. It supports cities and communities worldwide in addressing urban challenges such as housing, infrastructure, water, sanitation, and climate resilience. https://unhabitat.org/. Retrieved 20 November 2024.

25 Hyperbioregionalism builds upon bioregionalism, which advocates for living and designing in accordance with the ecological and cultural characteristics of a specific geographical region, including its climate, topography, biodiversity, and traditions (McGinnis, 1999). Bioregionalism emphasises place-based solutions informed by local knowledge systems to foster sustainable practices. Hyperbioregionalism expands this by integrating digital technologies, systems thinking, and a planetary ecological awareness and promoting design strategies that are simultaneously context-sensitive and globally informed.

References

Abbott, H. (2008). *The Cambridge introduction to narrative.* Cambridge University Press.

Brossmann, J., & Islar, M. (2020). Living degrowth? Investigating degrowth practices through performative methods. *Sustainability Science*, 15, 917–930.

Cattermole, J., Halba, H., Kalogeropoulou, S., & Tucker, H. (Eds.) (2021). Performing ecologies. *The Performance of the Real*, 2 (Special Issue).

Ceballos-Lascurain, H. (1996). *Tourism, ecotourism, and protected areas: The state of nature-based tourism around the world and guidelines for its development.* IUCN Publications. 301 pp.

Cheng, P. (2024). Basic principles and clinical aspects of Taoism. In H. G. Koenig, S. Meador, & J. G. Peteet (Eds.), *Eastern religions, spirituality, and psychiatry* (pp. 167–174).

Endicott, K. (Ed.). (2016). *Malaysia's original people: Past, present and future of the Orang Asli.* NUS Singapore.

Fry, T. (2022). *Writing design fiction: Relocating a city in crisis.* Bloomsbury.

Gellner, D. (1986). Language, caste, religion and territory: Newar identity ancient and modern. *European Journal of Sociology*, 27 (1), 102–148.

Gergen, M., & Gergen, K. J. (2010). Performative social science and psychology. *Qualitative Social Research.* 12(1).

Heras, M., & Tàbara, J. (2014). Let's play transformations! Performative methods for sustainability. *Sustainability Science*, 9. 379–398.

Hughes, R., & Huby, M. (2002) The application of vignettes in social and nursing research. *Journal of Advanced Nursing*, 37, 382–386.

Hughes, R., & Huby, M. (2004). The construction and interpretation of vignettes in social research. *Social Work and Social Sciences Review*, 11(1). 36–51.

Khadka, A., Khadka, A., & Parajuli, A. (2022). Municipal solid waste and its implication in the 2022 cholera epidemic in Kathmandu. *Patan Prospective Journal*, 2, 1. doi:10.3126/ppj.v2i1.48009.

Kant, I. (1998). *Critique of pure reason.* The Cambridge edition of the works of Immanuel Kant. Cambridge University Press.

Keitsch, M. (2015). Boundary objects as means for knowledge generation in design education, International Conference on Engineering and Product Design Education, 3–4 September 2015, University of Loughborough, United Kingdom. Design Society. ISBN: 978-971-904670-904636-0

Lucie-Smith, E. (2020). *Movements in art since 1945*. Thames & Hudson.

McGinnis, M. V. (Ed.). (1999). *Bioregionalism.* Routledge.

Østern, T. P., JusISCn, S., Nødtvedt Knudsen, K., Maapalo, P., & Bjørkøy, I. (2023). A performative paradigm for post-qualitative inquiry. *Qualitative Research*, 23(2), 272–289.

Perkins, D. N., Farady, M., & Bushey, B. (1991). Everyday reasoning and the roots of intelligence. In J. F. Voss, D. N. Perkins, & J. W. Segal (Eds.), *Informal reasoning and education.* Lawrence Erlbaum Associates. pp. 83–106.

Salvatore, S., & Valsiner, J. (2010). Between the general and the unique: Overcoming the nomothetic versus idiographic opposition. *Theory & Psychology*, 20 (6), 817–833.

Schweitzer, A. (1933). *Out of my life and thought: An autobiography* (C. T. Campion, Trans.). Henry Holt. (pp. 116–117).

Scott Hanson, C., & Scott Hanson, K. (2005). *The cohousing handbook: Building a place for community* (2nd rev. ed.). New Society Publishers.

Smith, N. (1984). *Uneven development: Nature, capital and the production of space.* Blackwell.

Singh, B., & Keitsch, M. M. (2025). Reflective agency: The practitioner's role in value-informed design practice. *The International Journal of Design Education*, 20 (1), 239–264.

Singh, S. (2018). Elements of eco city in traditional settlements of Kathmandu Valley. Doctoral dissertation, Department of Architecture, Institute of Engineering, Tribhuvan University. Lalitpur, Nepal.

Thrift, N. (2007). *Non-Representational theory: Space, politics, affect* (1st ed.). Routledge.

UNESCO. (2011). *Recommendation on the Historic Urban Landscape.* United Nations Educational, Scientific and Cultural Organization. https://whc.unesco.org/en/activities/638/. Retrieved 20 May 2024.

United Nations Office for Disaster Risk Reduction. (2015). Sendai Framework for Disaster Risk Reduction 2015–2030. https://www.undrr.org/publication/sendai-framework-disaster-risk-reduction-2015-2030. Retrieved 20 May 2024.

United Nations. (2017). *New Urban Agenda.* United Nations Human Settlements Programme (UN-Habitat). https://habitat3.org/the-new-urban-agenda/. Retrieved 20 May 2024.

Van Manen, M., & Van Manen M. (2021). Doing phenomenological research and writing. *Qualitative Health Research*, 31(6), 1069–1082.

Varto, J. (2011). *Song of the earth: Lectures on ethics.* Aalto University School of Art and Design, Department of Art.

Windelband, W. (1894), *Geschichte und Naturwissenschaft*. Rede zum Antritt des Rektorats der Kaiser-Wilhelms-Universität Straßburg gehalten am 1. Mai 1894. https://www.hs-augburg.de/~harsch/germanica/Chronologie/19Jh/Windelband/win_rede.html. Retrieved 20 May 2025.

wordreference.com (2024), *Dictionary of English*, idio - (graph).WordReference.com. Retrieved 20 May 2024.

Østern, T. P., JusISCn, S., Nødtvedt Knudsen, K., Maapalo, P., & Bjørkøy, I. (2023). A performative paradigm for post-qualitative inquiry. *Qualitative Research*, 23(2), 272–289.

5 Environmental Ethics-Inspired Pathways in Sustainable Design

Perspectives for Research and Education

5.1 BioDialectics – An Integrative Ethics for Sustainable Design

Biocentrism values all life forms and processes and can motivate ecological awareness and environmental responsibility. However, as previously discussed, biocentrism as a philosophical concept is not unproblematic, implying contradictions, ethical dilemmas, and socio-political challenges such as the risk of eco-fundamentalism.

Anthropocentrism regards humans as the sole moral agents, responsible for making and enacting moral judgments. As animal rationale, human beings possess the capability for rational thought and reflection, enabling them to plan and make ethical decisions about their impact on the natural world. However, despite this capability, human actions often cause environmental degradation, climate change, and ecological imbalance. In my view, this is partly due to an overreliance on instrumental rationality – a type of reasoning focused on achieving specific ends through efficient, goal-oriented thinking. While instrumental rationality can be useful, particularly in areas that value efficiency and calculability, its dominance often sidelines ethical and moral concerns. For example, it can lead to viewing nature merely as a resource to be efficiently and even "sustainably" exploited. In public policy and governance this can grow into strategic rationality, where decision-making is driven solely by utility, often ignoring broader ethical implications. As Habermas (1987) notes, strategic rationality can produce far-reaching social mischiefs, including pervasive mass surveillance, bloated bureaucratic systems, and environmentally destructive decision-making.

Instrumental rationality tends to focus only on the external value of nature and what it can provide to meet human goals. However, this implies a limited perspective on the environment, which, to account for the various aspects of nature, should be supplemented by concepts that recognise the inherent (or internal) value of nature. As Ariansen (1997) explains, inherent values can arise from subjective experience or from objective assessments of what supports the well-being of nature and its inhabitants, for example, a moose preferring a forested habitat. An integrative ethical framework should also consider the inherent (or internal) value of nature.

DOI: 10.4324/9781003384090-5

The BioDialectics approach, introduced in the following section, draws on ethics, ecology, aesthetics, and relational theory to inform alternative conceptual pathways in sustainable design and environmental ethics and -philosophy. It condenses insights presented in this book and should be understood as a propaedeutic introduction offering orientation that, at this stage, remains exploratory. It outlines core dimensions without claiming theoretical completeness, leaving space for further development and interdisciplinary dialogue.

The BioDialectics approach draws on and modifies Aristotle's triad of *theoria* (contemplation), *praxis* (ethical action), and *poiesis* (creative production). This triad is still employed in contemporary ethical and educational discourse to structure human reflection, engagement, and practice. York et al. (2012) and Makrakis et al. (2021) argue, for instance, that distinguishing among contemplative understanding, ethical action, and creative production enables a more holistic and transformative approach to ethical learning.

The dimensions of BioDialectics link the triad to embodied recognition, ethical awareness, and productive tension. The first dimension, *theoria*, draws on a Hegelian dialectical framework to reframe nature and humanity as interdependent entities engaged in a continuous process of encounter, contradiction, and resolution. However, unlike classical models of knowledge rooted in universality or linear progression, BioDialectics expands this into a pluralistic epistemology. Knowledge about nature is not derived solely from scientific reasoning but also through cultural, emotional, and ecological ways of knowing. This approach aligns with concerns of epistemic equity and justice by affirming the legitimacy of multiple knowledge systems, including Indigenous perspectives, embodied knowledge, and aesthetic insight rather than privileging technocratic or exclusively objective paradigms. While Araki (2021) identifies four components of knowledge – namely theoretical knowledge, practical skill, moral sensitivity, and reflective awareness – BioDialectics expands this framework by introducing the concept of disruptive knowledge: insights that challenge one's own and others' worldviews or unsettle (anthropocentric) assumptions[1]. Rather than avoiding contradiction, BioDialectics embraces cognitive dissonance as a generative force in epistemic development. In this view, *theoria* becomes contemplative, critical, and co-generative, allowing diverse interpretations of human–nature relationships to coexist.

Theoria is further connected to heuristic reasoning, in which the intrinsic value of nature is interpreted as if it was empirically knowable. Heuristics do not seek verification but instead invites reflective engagement. For example, critical anthropomorphism enables the imaginative perception of nonhuman lifeworlds, creating space for both epistemic and ethical reconsideration. Heuristics also serve to formulate alternative concepts and to bridge *theoria* and *praxis* via ethical imagination without prescribing moral actions, and by cultivating epistemic humility that grounds ethical standpoints in relational understanding.

The second dimension, *praxis*, relates to the socially mediated interpretation and communication of nature encounters, and to how values, narratives, and cultural frameworks influence human–nature relationships. In this sense, *praxis* reflects the interplay between human agency and environmental surroundings, which may involve ethical values such as respect, autonomy, and non-identity[2]. These values can lead to moral action, even though they are not framed as prescriptive moral imperatives. Instead, BioDialectics adopts a view of ethical responsiveness as emerging from the interplay of emotional, rational, and lived experience within pluralistic, context-dependent settings. This view reflects a broader shift in contemporary discourse and has gained increasing legitimacy in fields such as sustainable planning and design. For example, Fernandes et al. (2016) argue for rethinking environmental ethics as a situated practice embedded in emotion, identity, and social meaning. Similarly, Barrett et al. (2016) and Osborne et al. (2015) emphasise the importance of integrating emotional awareness, ethical reflection, and lived experience into urban planning processes, highlighting the relevance of this approach beyond philosophical contexts.

The first mode of the third dimension, *poesis,* refers to the aesthetic and sensory experience of nature, rooted in aisthesis, i.e., perception through the senses. Embodied and perceptual engagement is considered central to the cultivation of environmental awareness[3]. This mode underlines direct encounters with other beings in shared environments, with particular emphasis on liminal experiences, where boundaries between self and other dissolve. Intersubjective *poesis* may contribute to questioning established moral frameworks and suggesting alternative ways of relating across species and natural systems. The second mode of *poesis*, as productive activity, can be aligned with the term "productive paradox"[4], coined by the author of this book. Productive paradox refers to the generative power of holding unresolved fissures and tensions, between knowing and ignorance, control and submission, self and world. Rather than resolving contradictions, productive paradox represents a poietic method for generating inspiration, insight, and ethical sensitivity rendering *poiesis* a studio of contingency.

In all dimensions, BioDialectics necessitates a departure from traditional biocentrism's emphasis on the intrinsic value of all living beings, no matter what, to a broader concept that incorporates the relational and systemic dimensions of the environment by addressing historical, geo-political, and structural inequalities that shape it. Justice, in this concept, expands human fairness to one another to ecological decolonialisation by equally addressing burdens placed on marginalised communities, ethnicities, and ecosystems through environmental degradation and resource extraction. Embedding justice at the core of BioDialectics, particularly through its cognitive and emotional dimensions, may lead to more equitable visions of environmental stewardship.

Explanations of reciprocities between human and non-human beings begin at a meta-cognitive level, namely with understanding how anthropomorphism influences human decision-making. Rather than treating anthropomorphism as a mere cognitive distortion or bias, BioDialectics deliberately foregrounds it as a generative and reflective lens through which to understand nature, its attributed values, and the ethical responsibilities such perceptions entail. In my opinion, the apperceptive capacity represents one of the most distinctive features of the *conditio humana* and represents the starting point in BioDialectics for engaging with the ethical dimensions of human–nature relationships. By engaging two distinct dimensions of anthropomorphism, BioDialectics seeks to deepen experiential perception, foster emotional engagement, and promote ethical reflection on humanity's interconnectedness with the natural world insights that are particularly relevant to the responsibilities and practices of the sustainable designer.

As the narratives in Chapter 4 illustrate, naïve projections of human characteristics onto nature appear in various contexts, often reflecting unexamined or intuitive responses. In contrast, systematic and critical anthropomorphism adopt a more deliberate approach, seeking to bridge subjective experience with empirical insight and ethical awareness. Within this lens, interpretations of nature encompass a range of aspects, from scientific understanding to cultural narratives, social settings, and subjective aesthetic experiences.

Guthrie (1993) plays a significant role in discussions of systematic anthropomorphism. He argues that humans perceive their environment through an anthropomorphic lens that leads to the humanisation of nature. This tendency is reinforced through rituals, symbolic representations, and mythological frameworks across cultures. Guthrie draws on various examples from the history of philosophy to support his argument:

> *"On the continent, the encyclopedist Denis Diderot imagined that 'the world, like a huge animal, has a soul [which] may be an infinite system of perceptions' extending throughout matter … Diderot's contemporary, J. B. R. Robinet, wrote, 'All matter is organic, living and animal … This invisible world [upon which matter is based] is the aggregate of all the forces which continually strive to ameliorate their existence …' Similar animism existed in Germany, most notably among the Romantics, and elsewhere in Europe. Despite Descartes and Newton, many Europeans continued to believe that 'all matter was living, organic, and animal, that all natural objects, as organised forms of matter, had their own life and sensibility, and that the whole organisation of the natural world was capable of intelligent purpose'".*
>
> (Guthrie, 1993, 58)

Silva et al. (2025) interpret Guthrie's systematic anthropomorphism as grounded in evolutionary phylogenesis. It encourages reflexive understanding of human interconnectedness with the natural world. For Guthrie, anthropomorphism is not a mere cognitive bias but an evolutionary byproduct shaping how humans perceive agency, relate to non-human entities, and potentially develop ethical commitments to the environment.

Ethically, Guthrie's position resonates with Spinoza's notion of intrinsicality, where ethical relations to nature do originate in human autonomy or moral structures but from an understanding of necessity. Seeing nature as relational other rather than as a passive backdrop to human action implies that environmental destruction is not only ecologically harmful but also ethically wrong.

Perception, valuation, and protection are closely intertwined in Guthrie's concept. He suggests that interpreting nature as a relational partner is a lived reality in many Indigenous and religious traditions, which approach landscapes, plants, animals, and ecosystems with reverence and regard them as entities possessing spiritual or intrinsic worth:

> *"The Iban, for instance, identify the spirits of rice with those of humans. Rice 'is just like one of us' and is accorded the utmost reverence and respect … [It has] human moods and attitudes: it is 'unhappy', 'feels unwanted' – Iban women 'take pity' on small grains while harvesting – it 'catches cold', 'needs company', and 'likes attention'".*
>
> (1993, 113)[5]

Before turning to Glissant's concept of critical anthropomorphism and decolonialisation, along with its implications for nature–culture encounters, the following section examines key challenges within Guthrie's conceptual approach.

Logically, experiences and perceptions of nature and ethical values are related through correspondence rather than direct causation, as articulated in Hume's is-ought problem (1739), which holds that prescriptive statements (ought) cannot be directly derived from descriptive statements (is). A more detailed discussion of the is-ought problem is provided in Chapter 2.1.

Phenomenologically, systematic anthropomorphism may overlook aspects of nature's non-identity and openness, as well as its independent existence (especially as *natura naturans*), which surpasses human perception, and leaving nature experiences inherently indeterminate and open-ended. Socially and symbolically, interpretations of nature inform and enrich cultural narratives, but they can sometimes conflict with scientific perspectives on ecology. Disregarding ecological science can lead to a backfire effect, where the oversimplification or dismissal of empirical data hinders environmental action.

Some views of nature also may become dominant and support political ideologies and power structures. A historical example of the ideologisation of human phylogenies is the social hygiene discourse of the early twentieth century, in which notions such as "healthy (natural)" and "unhealthy (unnatural)" increasingly replaced moral categories of "right" and "wrong", thereby transferring ethical judgment into a biological framework (Keitsch, 2002). Meanings attributed to nature are hence not neutral but are shaped by historical contexts, social hierarchies, and ideological interests that influence how nature is represented, governed, and ethically valued.

Eventually, cultural habits within late-capitalist consumer societies can reduce nature perceptions and experiences to a Sonntagsvergnügen ("Sunday entertainment"), transforming them into episodic, recreational activities.

According to Sperb (2016) and Cronon (1996) this process reinforces anthropocentric power structures by transforming non-human entities into experiential commodities, where nature is primarily valued for providing relaxation, and pleasure rather than being recognised as an autonomous or relational other.

Glissant's (1997) concept of critical anthropomorphism is socio-politically framed, exploring the intersections of nature, identity, colonial histories, and political legacies. His decolonial approach links landscapes to cultural memory and resistance, challenging dominant Western-centric narratives of nature. Central to this framework are the concepts of opacity and the poetics of landscape. Opacity resists dualisms between nature and culture, human and non-human, and subject and object. Beings are understood as constituted through relations (Glissant, 1997, 144), continuously emerging and dissolving rather than existing as fixed, self-contained entities. Opacity thus emphasises fluidity and polyvalence while rejecting complete transparency, instead advocating openness to what remains non-identifiable. In this way, landscapes, oceans, and ecosystems become active participants in historical and cultural processes rather than passive backdrops to human history.

Glissant's understanding of landscape offers an evocative and thought-provoking articulation of critical anthropomorphism, with emphasis on interaction and entanglement between humans and the more-than-human world. Landscape is not conceived as a mere physical assemblage or static scene but as a palimpsestic space that carries history, memory, and identity. In the Caribbean context, landscapes embody displacement, colonial violence, and resistance, shaped by Indigenous, African, and European actors alike. As Glissant writes,

> *"Crossed and elevated by the trace, the landscape ceases to be a suitable décor and becomes a part of the drama of Relation. It is no longer the passive outer layer of the omnipotent Narrative but the changing and unnecessary dimension of all change and exchange".*
>
> (2005, 21; 2007, 18)

Glissant develops a novel relational ontology of nature. Yet he also constructs a fundamental opposition to Eurocentric philosophy, thereby reducing the diversity of nature concepts within it and overlooking authors such as Heidegger and Adorno, who seek to relate human–nature antinomies. Heidegger, for example, draws on the Aristotelian notion of *peri ti* ("around-this") (Keitsch, 2003). Following this line of thought, a possible philosophical rapprochement between Glissant's relational and Heidegger's fundamental ontology can be articulated through Aristotle's notion of *tode ti* ("some this"). Rather than denoting a fully determined substance, *tode ti* refers to an entity whose existence unfolds through evolutionary and culturally shaped processes and resists exhaustive definition (Chen, 1957). It thus emerges as a shared philosophical point of departure for thinking about existence, openness, and non-identity. The notion aligns with Heidegger's efforts to transcend human–nature antinomies and resonates with

Glissant's emphasis on becoming and his resistance to totalisation, a position he shares with Adorno. Both embrace the fluidity of existence, Adorno as "non-identity" (1997) Glissant as "opacity" and both advocate for the irreducibility of subjectivity, culture, and nature.

Glissant's reflections on landscape can be read as a counterbalance to instrumentalised, commodified, and romanticised interpretations of nature. However, his limited engagement with counter-currents within European intellectual and artistic traditions risks oversimplifying philosophical concepts, neglecting internal critiques and poetic deviations from dominant paradigms, and overlooking potential affinities with European thinkers who similarly challenge fixed conceptions of nature and human identity. To avoid a new form of reductionism, Western frameworks should be approached in a more differentiated manner, acknowledging both their hegemonic tendencies and their critical counter-streams, such as phenomenology and postmodernism. An epistemic tension also present in post-anthropocentric approaches emerges from Glissant's assumption of the opacity of beings alongside his rejection of rational (Eurocentric) categorisation. If, however, conception (rationalisation) and perception (experience) are inseparably intertwined, perception itself is shaped by intellectual structures, thereby complicating claims of an entirely opaque mode of existence.

Setting aside philosophical intricacies, critical anthropomorphism plays a crucial role in reshaping human perspectives on nature by challenging rigid boundaries between the human and non-human. This resonates with the dialectical approach in BioDialectics, which similarly emphasises relationality, fluid identity, and openness to more-than-human agency.

Conclusively, BioDialectics provides a reflective, relational, and ethically responsive supplement to both environmental ethics and sustainable design. While not a fully developed concept, it serves as a philosophical entry point for further exploration. In the context of sustainable design, BioDialectics aligns with posthumanist and "more-than-human" perspectives already emerging in speculative and critical design discourse. It captures the complexity of human–nature relationships, where material, social, ecological, and symbolic dimensions intersect in dynamic and often contradictory ways.

Designers must continuously navigate between function, values, and context, and BioDialectics supports this process by illuminating these interdependencies rather than simplifying them. Its emphasis on embodied and experiential knowledge is particularly relevant in sustainable design, where cultural identity, belonging, and aesthetic values shape both human well-being and environmental action. By valuing such knowledge, BioDialectics opens space for human-centred, place-sensitive, and ethically grounded design practices.

Ethically, BioDialectics emerges in the interplay and tension between situated ethics and the deontic principle of justice. I would advocate for coupling a universal ethics of duty in the Rawlsian tradition with an ethics of responsibility cultivated through experiential engagement with and perception of nature. The deontological approach ensures the consistent application of interspecies or ecological justice. Ecological fairness provides a stable and coherent ethical framework by upholding obligations universally. Complementing this, a consequentialist ethics of responsibility responsive to lived experience addresses the contextual relationships between human and non-human beings within a shared lifeworld.

Eudaimonistic ethics, which holds that the highest ethical aim is the good life through the cultivation of virtue, also offers promising possibilities when extended beyond the human realm to include ecosystemic flourishing. This perspective recognises that human well-being is profoundly interconnected with the natural world and encourages virtues such as care, responsibility, and ecological humility (Küpers, 2020). Furthermore, relational ethics aligns with the opacity and polyvalence of real-life contexts, where values evolve dynamically (Keitsch et al., 2020). It emphasises embodied knowledge, spatial context, and intersubjective processes as essential conditions for realising ethical values. Bergum et al. (2005) describe for instance landscape as "ethical space" or "relational space", arguing that it must be recognised and nurtured to foster meaningful communication between human and non-human agents.

The ethical positioning of BioDialectics remains open to further philosophical scrutiny, yet the concept provides a credible onset for reinterpreting sustainable design. It challenges designers to engage more deeply and critically with the ecological and ethical dimensions of their work and offers a transformative approach on how they can respond to the complex and rapidly changing demands of environmental realities.

5.2 Felicities and Fallacies of Adopting Post-Anthropocentrism in Sustainable Design

In contemporary sustainability discourse, post-anthropocentric concepts such as BioDialectics are increasingly influencing traditionally human-centred nature interpretations. This chapter explores both the *felicities* (potential benefits) and *fallacies* (inherent challenges) of integrating post-anthropocentric thinking into sustainable design. It considers how ideas might broaden ethical awareness beyond human interests, while also critically examining the conceptual ambiguities and practical constraints that accompany such an approach.

Post-anthropocentric reflections challenge anthropocentrism not only on moral and ethical grounds but also by interrogating foundational research paradigms, which, as Lakatos (1970) argues, constitute the core of a scientific discipline. In the field of design, Simon's (1996) conception of design as a cognitive process for solving ill-structured problems represents such a paradigm. Post-anthropocentric thinking, for example, destabilises this paradigm by redirecting focus from human-defined problem spaces to more-than-human actor-networks. This shift implies questioning the assumption that design problems can be clearly defined, fully understood, and effectively solved from an exclusively human standpoint. Epistemically, post-anthropocentrism may hence pose significant challenges for design theorists, educators, and practitioners alike.

In terms of design practice, some post-anthropocentric concepts align with a regenerative perspective that supports ecological resilience and long-term sustainability. In this context, Mang and Reed's (2012) example of "Living System Design" interprets design as an evolving, participatory process embedded within ecological and social systems (see also Reed, 2007; Gibbons, 2020). The primary objective of design is here to enhance and restore ecosystems' capacity for self-organisation and resilience, ensuring their long-term sustainability and ability to thrive. The principle of interdependence between human and non-human systems is fundamental to implementing regenerative and Living System Design practices (Mang et al., 2012), and within this principle, sustainable design shifts focus from merely mitigating environmental harm to actively restoring and enhancing ecological systems. On a technical level, regenerative thinking comprises mimicking natural ecosystems for example, in buildings that self-produce energy, collect and treat rainwater for reuse, and process waste on-site through a composting system.

On a socio-political level, exploitative practices embedded within design methodologies are critically analysed and restructured to incorporate more equitable approaches encouraging eco-social transformation: *"Eco-social transformation, understood from the standpoint of regenerative practice, is centrally built on the regeneration of relationships with the embodied self, with other humans, and with more-than-human nature"* (Or, 2021, 229).

Epistemologically and institutionally, regenerative concepts tend to advocate for interdisciplinary collaboration among environmental sciences, philosophy, and design. Paolini et al. (2024) assert, for example, that the fluidity of regenerative concepts enables interdisciplinary adaptation while simultaneously upholding the principles of more-than-human environmental justice.

However, several authors also acknowledge that moving toward regenerative, post-anthropocentric, and just societies presents significant challenges for design. Among other things, when established epistemologies are questioned, this necessitates the development of novel frameworks and methodologies in both theory and practice (Borthwick et al., 2022). Further, conventional methods are not designed for regenerative or life-centric approaches, and in practice, professionals often lack the necessary training and knowledge to implement them effectively. Additionally, assessing the success of such designs requires supplementing conventional performance measures with metrics that account for multispecies justice (Zavoleas et al., 2023).

Shifting attitudes and values appear to be only the first step in developing new frameworks that integrate non-human agency into design processes (Bennett et al., 2016; Camocini et al., 2021). Beyond new knowledge generation and feasibility for implementation, post-anthropocentric design transformations need strong institutional commitment and supportive education policies (Raworth, 2017). Without comprehensive systemic support, the integration of post-anthropocentric ideas is likely to remain confined to experimental or peripheral initiatives, rather than becoming widely adopted within sustainable design. Transformative changes across governance, education, and industry are also crucial to securing both the viability and broader impact of such approaches (Gibbons, 2020).

Despite these challenges, post-anthropocentrism presents an epistemic and ethical opportunity to redefine sustainability in design beyond merely mitigating environmental degradation to actively regenerating relationships between ecosystems and social systems. Yet, pursuing a post-anthropocentric approach, in this book as BioDialectics, is not a straightforward endeavour. The next section explores the crucial role of educating designers to engage in post-anthropocentrism in sustainable design. It highlights the necessity of fostering critical thinking, interdisciplinary collaboration, and ethical awareness to actively contribute to the evolution of sustainable and regenerative design practices.

5.3 Shaping the Future of Sustainable Design Education

By integrating post-anthropocentric principles into sustainable design education, future professionals can develop the skills and perspectives needed to create more inclusive, ecologically responsible, and ethically grounded design solutions. Various publications and interviews with sustainable design educators, featured in the narratives, underline that imparting a sense of responsibility to students is an important task in design curricula. One educator connected the concept of responsibility to "design agency" in an interview, explaining that designers play a crucial role in ensuring the inclusion of all users and stakeholders in a project. By recognising and valuing stakeholders' contributions, more equitable and effective solutions that resonate with a wider audience can be co-created. When asked about the extent to which nature is part of design agency, the educator replied: *"The term 'nature' is too vast and abstract for me. I prefer to teach students to focus on context-sensitive design, integrating cultural, social, and environmental aspects into their concepts"* (Interview, narrative 4.2.2). The inherent modesty in this pragmatic approach is, in itself, ethically sound, and teaching students to incorporate ecological concerns not merely as a functional requirement but as an integral component of every design assignment is both laudable and convincing. However, design comprises many trade-offs, and when making selections and compromises, the environment is often the first stakeholder to be opted out. Practising the fundamental steps of an ethical analysis can, in this context, not only enhance design students' awareness of moral dilemmas but also reinforce their sense of professional responsibility, particularly in the development of sustainable solutions. Discussing ethics in sustainable design education can comprise both an analysis of challenges and a discussion of concepts such as participatory design with regard to their ethical implications.

An ethical analysis facilitates the evaluation and discussion of moral challenges, responsibilities, and shared norms. The significance of the discussion depends in turn on meeting certain conditions. First, a problem has to be recognised as a moral challenge. This recognition entails acknowledging both one's own and others' autonomy, for example in terms of thinking independently, free from external pressures, and engaging in reasoned argumentation, such as critically examining why the issue constitutes a moral challenge and what it encompasses. From a posthumanist perspective, the non-human can be regarded *as if* it has autonomy, allowing them to be included in moral considerations.

Second, ethical norms have to be justified by clearly determining what is permitted or prohibited and why. Justification has to be transparent, open to discussion, and defensible within a (discourse) community. For a norm to be considered valid, it must be widely accepted; a norm recognised by only a few individuals lacks legitimacy. Validation implies that a norm is broadly acknowledged as legitimate to ensure its acceptance.

An ethical analysis of meat consumption in the context of sustainable design education can, for instance, start with recognising ethical consumption as a significant moral challenge. This recognition involves a critical examination of various aspects such as animal welfare, environmental degradation, health considerations, personal well-being, individual choices, and the freedom to adopt different lifestyles. Justification considers different arguments such as the suffering of animals in industrial farming, the depletion of natural resources, individual autonomy, the cultural, historical, and religious significance of meat, the livelihoods of farmers, and the potential for sustainable meat production, to determine what is permissible, justifiable, and widely acceptable within society. Justification can further include strategies such as evaluating whether, for example, government policies restricting meat consumption, similar to regulations on alcohol or drugs, could be ethically justified in protecting public health and the environment, or whether such measures would constitute an infringement on personal autonomy. Additionally, discussions on strategies should consider how to ensure public acceptance, i.e., that norms align with societal values and are broadly acceptable within a community.

Sustainable design curricula can integrate ethical considerations to varying degrees, and understanding the fundamental conditions for ethical analysis can support the development of more integrative and balanced solutions. While ethics provide a common ground for defining what ought to be done, Participatory Design (PD) can serve as a practical framework in sustainable design education, bridging ethical reasoning with socially and ecologically responsible action. PD offers methods to ensure inclusivity, transparency, and legitimacy throughout the ideation, design, and implementation phases. Traditionally, PD approaches emphasise the importance of understanding diverse stakeholder perspectives, acknowledging their differing interpretations, needs, and skills (Sanders et al., 2008). Through co-creation, ethical norms evolve collaboratively rather than being imposed, with experts facilitating rather than imposing decisions. Shared ownership enhances both the legitimacy and effectiveness of decisions, as inclusive participation fosters stakeholder commitment and reinforces social responsibility (Zamenopoulos et al., 2018).

Recently, multispecies design and speculative participatory design (SPD) approaches have gained ground in the sustainable design education debate. Gatto et al. (2019) propose, for instance, a framework that extends participation to plants, ecosystems, and microbial life. Multispecies design interprets design as *"a process-oriented critical instrument"* (Gatto et al., 2019, 2) rather than a strictly defined field of practice. The concept integrates ethnographic methods, post-humanist theories, and speculative participatory design (SPD), advocating for designing *with* rather than *for* the environment. Methodologically, it proposes embodied knowledge and (critical) anthropomorphic explorations to foster engagement with non-human perspectives and interspecies relationships. A valuable supplement to SPD is the inclusion of the aesthetic experience of nature through hermeneutic phenomenology, as proposed by e.g. Iared et al. (2016). Drawing on Gadamer's hermeneutics and Merleau-Ponty's phenomenology, they argue that nature experiences foster profound, embodied, and emotional engagement with other beings. Informed by Freirean pedagogy[6], Iared et al. interpret these encounters as critically reflective, encouraging learners to question dominant narratives and cultivate ethical, relational understandings of nature. In education, both SPD and hermeneutic approaches invite students to connect their own lived experiences with observations and insights from various sources – including those concerning non-human beings – thus generating "situated knowledge".

This "idiographic knowledge", though complex and sometimes contradictory, expands understanding of how lived experiences shape choices and actions. For example, a student exploring meat consumption can document their own dietary choices, and observe different cultural, ecological, and social conditions and habits that influence attitudes towards meat. Visiting both industrial and regenerative livestock farms and experiencing how animals are kept, engaging with plant-based dietary movements, and consulting health experts, they gain insights into the interconnections between human consumption habits, the well-being of animals, and the impacts on ecosystems and local communities. In this embodied approach, "situated knowledge" helps move beyond arguments over meat versus plant-based diets, ethics, and ecological responsibility.

The embodied approach in multispecies design enriches education by fostering context-aware, experience-driven learning. It enables students to integrate individual lifeworld experiences with design, expert perspectives, products, services, and non-human elements, generating new insights into socio-environmental conditions, materials, and ecosystems. However, excessive subjectivity, emotional and eco-political bias, and deviation from conventional sustainable design models may hinder the adoption of multispecies design. Accordingly, integrating this post-anthropocentric concept into curricula necessitates critical anthropomorphism as an essential complement.

Speculative participatory design recognises the agency of non-human entities, such as plants, animals, and ecosystems, in shaping environmental and technological systems. Challenging anthropocentric design paradigms and advocating for ethical and ecological considerations that transcend human interests (Gatto and McCardle, 2019; Jacobs et al., 2025), SPD integrates participatory methods with speculative design for inclusive, future-oriented thinking. The aim is not to predict the future but rather to create participatory spaces where alternative possibilities can be debated and explored. Engaging diverse stakeholders in imagining and shaping potential futures, SPD employs storytelling, prototyping, and performative interventions to materialise these futures, provoking critical discourse rather than prescribing fixed solutions. Inclusion encompasses non-human agents such as ecosystems, materials, and artefacts, recognising them as active participants in the design process (Giaccardi et al., 2020). As Wilkie et al. (2023) anthropomorphically state:

> *"To be human is to be more-than-human ... In every human activity, from the most individual to the most collective, the nonhuman is present: even in the depths of our innermost dreams we are embroiled with physical surfaces, chemical compounds, and a multitude of microbes".*
>
> (ibid., 105)

While the role of humans varies across different concepts, ranging from being "part of an entanglement", as illustrated above, to network-based approaches that integrate ecological ethics and non-human agencies (Hupkes and Hedman, 2022), SPD generally encourages educators to rethink pedagogical models. This involves adopting frameworks that emphasise ecological sustainability and the agency of materials or artefacts. For instance, educators might create co-creative scenarios with plants, bacteria, or autonomous AI systems to envision pluralistic futures (Nhemachena, 2024). This shift towards post-anthropocentrism in sustainable design education can broaden design perspectives by encouraging students to acknowledge non-human agents, such as ecosystems, materials, and artificial intelligence, as participants in the design process. Additionally, post-anthropocentric sustainability awareness helps future designers to include ecological and ethical considerations in design trade-offs. Speculative design facilitates the development of critical thinking, which might include a stronger sense of ethical responsibility, encouraging reflection on the impacts of solutions beyond human users.

In summary, ethical decision-making ensures that norms are rational, justifiable, and inclusive, but ethical principles must be translated into real-world actions. By challenging the assumption that human experience is the primary reference point for design and embracing ambiguity, SPD can act as a bridge between ethical theory and practical implementation. SPD prepares students for emerging design challenges, such as climate change, AI ethics, and post-human interactions, by equipping them with the skills needed to navigate these complexities and can contribute to the development of policies and solutions that are collaborative, transparent, and reflective of shared values. Besides all these positive aspects, however, a crucial question remains: How can a predominantly nomothetic epistemology in sustainable design be adapted, transformed, or even replaced to integrate these new mindsets into sustainable design education?

5.3.1 Excursion: Sustainable Design Epistemologies

Sustainable design oscillates within an epistemological dialectics of universal principles and localised adaptation by relying on two epistemologies: the nomothetic and idiographic paradigm (Singh and Keitsch 2025). Nomothetic epistemologies dominate the discipline since design transformed into an academic field with a focus on universal principles, standardisation, and quantitative methods. As Simon famously stated: *"Engineers are not the only professional problem solvers. Everyone designs who devises courses of action aimed at changing existing situations into preferred ones"* (1996, 111). Idiographic epistemologies, on the other hand, emphasise contextual specificity, qualitative insights, and localised knowledge.

From a philosophy of science perspective, the nomothetic approach in sustainable design is grounded in scientific rationalism, positivism, and empiricism. The aim is to formulate generalisable principles that surpass contextual variability and that can be systematically applied in diverse settings. Methodologically, quantifiable metrics, algorithms, and predictive modelling are translated into design tools such as Life Cycle Assessment (LCA) and Material Flow Analysis (MFA) and inform standard frameworks like Leadership in Energy and Environmental Design (LEED) and Sustainability Indicators (SI). Tools and frameworks, in turn, support circular economy models, drive efforts to reduce carbon footprints, and guide economic and policy decision-making, thereby shaping corporate strategies and sustainable governance structures.

While nomothetic epistemologies establish universal standards that are broadly applicable, they do not consider the complexity of localised ecological and socio-cultural contexts. In contrast, idiographic epistemologies reflect the particularities of place, culture, and individual experience by employing methods such as ethnographic research, participatory design, and co-creation, customising solutions for specific environments and communities. "Idiographic" sustainable design examples include, for example, Lo-Tek (Watson 2019) i.e., indigenous and nature-based architectural and design practices and frugal innovation (Mattos and Moreira, 2020) approaches. Acknowledging subjective and situated knowledge and ensuring that sustainable design solutions are responsive to local needs and environmental conditions, are main characteristics of applied idiographic knowledge integration. However, idiographic approaches can be difficult to scale and may struggle to fit within broader policy frameworks (Singh and Keitsch, 2025).

Beyond methodology, the epistemological divide in sustainable design also extends to ethics. Nomothetic ethics rely on universal principles such as human rights, ecological justice, and sustainability standards, aligning with deontological and utilitarian ethical frameworks. However, strict adherence to universal ethical norms may overlook cultural and ecological specificities, leading to ethical imperialism or moralism. In contrast, idiographic and living ethics (Kinunen and Valtonen, 2017) approaches emphasise context, relational responsibility, and the moral agency of both human and non-human actors. This aligns with virtue ethics, care ethics, and post-human ethics, recognising that ethical considerations must be adaptive to diverse realities. The challenge lies in developing a balanced ethical framework that honours both universal principles of sustainability and culturally specific moral perspectives. As discussed earlier, BioDialectics is a propaedeutic for developing such a framework.

Navigating epistemological and ethical tensions is challenging for sustainable design education. Traditional design curricula often favour nomothetic approaches, emphasising technical expertise and standardised methodologies. However, fostering a more integrated perspective requires acknowledging the value of idiographic knowledge systems. By incorporating participatory methods, localised case studies, and interdisciplinary approaches, sustainable design can become more inclusive, responsive, and effective, facilitating epistemic justice (Maldonado et al., 2023).

Conclusively, the integration of nomothetic and idiographic approaches (Hermans, 1988; Dieleman, 2008) offers a comprehensive knowledge framework for, among others, sustainable design education, which enables future designers to pair evidence-based methodologies, such as environmental impact assessment, with localised systems, including cultural practices, indigenous knowledge, and place-based experiences. Baskerville et al. (2015) argue that design science research already synthesises this pluralist epistemology. According to them, design-oriented research frequently starts idiographically, discussing a specific, localised problem, then implements nomothetic methods and abstracts insights by designing artefacts. Alternatively, the design process starts with generic theories and relates them to specific contexts, thus engaging in idiographic knowledge. Østern et al. (2023) even surpass the pluralist epistemology with their concept of a performative epistemic paradigm, which neither aligns with the nomothetic quest for general laws nor confines itself to idiographic depictions of the unique. Instead, they propose a radical reimagining of research as a creative, embodied, and ethical process that is always emergent and relational. Through this lens, knowledge is not found or constructed but enacted and performed, continuously reshaping the researcher, the phenomenon, and the world in which they are entangled. The performative paradigmatic shift

> *"liberates the researcher and ensures that it is acceptable, desirable and required to be embodied and affected, and that, as a researcher, one must not try to peel off sensory ways of communicating, understanding and engaging with a research phenomenon, or try to apply meaning-making articulations to written language only".*
>
> (Østern et al., 2023, 14)

Araki is another author who expands the understanding of professional expertise through a fourfold knowledge framework consisting of epistemic, tooled, idiosyncratic, and meta-knowledge (Araki, 2021, Singh and Keitsch, 2025). He challenges the theory–practice divide by emphasising that expert knowledge is multidimensional, situated, and reflexive. Building on Herling's (2000) theoretical–practical distinction, epistemic knowledge refers to formal, propositional understanding, while tooled knowledge involves the operational application of methods and tools.

Araki adds idiosyncratic knowledge, rooted in embodied experience and cultural context, as well as meta-knowledge, which reflects self-regulation and epistemic awareness. These diverse forms of knowing underscore the importance of epistemological plurality in design processes. Conclusively, by adopting multiple ways of knowing, sustainable design becomes both technically robust and socially resonant, fostering solutions that are not only sustainable but also contextually significant and resilient. An example is smart city planning that implements global metrics while adapting solutions to local cultural and ecological contexts, allowing for both environmentally resilient and socially inclusive forms of sustainable well-being.

5.4 Emerging Challenges in Sustainable Design Education

Contemporary sustainable design education is increasingly challenged to integrate intersecting dimensions, including global sustainable development policies, accelerating environmental degradation, technological innovation, and cultural appraisal. Within this context, performative design, post-anthropocentrism, decolonial thinking, novel materials and their interpretations, and artificial intelligence (AI), among others, are gaining increased attention in research and contributing to the development of new design methodologies, curricula, and professional practices.

Performative design expands the methodological landscape of sustainable design education by prioritising process over product and emphasising embodied experimentation, community engagement, and responsiveness to context (Østern et al., 2023). Such approaches allow students to explore environmental issues in real-time and real-place contexts, fostering a participatory understanding of ecological challenges (DiSalvo et al., 2022). Although performative methods are not always easily conducted, assessed, or preserved, they facilitate inclusive forms of learning that encompass recognition of existence, lived experience, relationality, reflection, action, and creative expression. Moreover, performative design accepts the contingency and "opacity"[7] of commonly generated solutions acknowledging that outcomes may remain open-ended and contextually shaped rather than predefined or easily measurable.

Post-anthropocentrism recentres the relationships between human and non-human actors, challenging established anthropocentric hierarchies. Ethically, it affirms the agency, value, and interdependence of all forms of life and matter. In sustainable design education, moving beyond user-centred approaches to include materials, species, and environments fosters systems thinking, multispecies perspectives, and regenerative practices. However, as previously discussed, concepts also present several challenges. While conceptually rich and intellectually compelling, post-anthropocentric approaches demand sustained critical deconstruction and the careful translation of abstract principles into practical, adaptable methodologies for teaching.

Decolonial concepts are increasingly challenging Western, anthropocentric foundations of sustainable design and promote diverse indigenous knowledge systems, relational ontologies, and plural ecologies (Whyte, 2017). In education, this shift contributes to include cultural specificity, land-based practices, and a focus on epistemic justice (Maldonado et al., 2023). Students are now invited to explore different ways of understanding sustainability and human–nature relations. However, an outright epistemological rejection of Western scientific knowledge risks dismissing valuable conventional research innovations and overlooking their potential contributions to sustainability merely on the basis of their origin (Koskinen et al., 2011). According to Appadurai (2013), some decolonial concepts also tend to equate all Western practices with colonialism, ignoring internal reform movements within Western science and design that actively seek collaboration with indigenous and local communities. Ethically, intentions in decolonialisation approaches are strong, while their methodological integration into teaching often remains vague, raising concerns that they may remain symbolic rather than lead to real changes in pedagogy or design practices (Tlostanova et al., 2012). Decolonial design thinking offers a critical framework for rethinking sustainability, while cooperation rather than opposition between diverse knowledge systems provides a viable, future-oriented trajectory for sustainable design.

Material innovation redefines another frontier of sustainable design. Innovations such as mycelium-based composites, algae bioplastics, and biodegradable textiles materials comprise reduced environmental footprints in compatibility with circular design principles (Myers, 2012). Many of these materials remain experimental in practice, rely on resource-intensive production processes, or raise concerns related to land use, labour, and socio-economic accessibility (Tsing, 2015). Yet, their integration into sustainable design education enables connections between technical exploration, aesthetic inquiry, and ecological evaluation, thereby reflecting interdisciplinary and transdisciplinary collaboration across design, science, engineering, and art (Collet, 2020).

Philosophically, new materials prompt a revision of agency. Several authors emphasise materials as active participants in the design process rather than treating them as inert substrates to be manipulated (Myers, 2012; Parisi et al., 2022). In educational contexts, this perspective encourages students to cultivate atunement to materials, in terms of their physical properties as well as regarding cultural, ecological, and political interpretations (Tsing, 2015). Engaging with materials in this way invites a reflective and situated design practice. The aesthetic language of emerging materials is particularly relevant to sustainable design, as biomaterials and experimental composites often display visual and tactile qualities that differ markedly from those of conventional industrial materials (Myers, 2012). Irregular appearances, soft textures, and variable, less rigid or predictable forms stand in contrast to traditional design aesthetics, which tend to privilege uniformity, stability, and control. In working with such materials, students encounter productive tensions between form, function, and environmental complexity. According to Ingold (2013), the aesthetic negotiation via material encounters reflects a broader shift towards embracing imperfection, adaptability, and ecological embeddedness in design (Ingold, 2013). This process not only challenges dominant design paradigms but also cultivates a more nuanced understanding of sustainability as a material, experiential, and aesthetic practice.

Finally, the integration of artificial intelligence (AI) into sustainable design education introduces new opportunities for simulating ecological outcomes, optimising material use, and managing complex systems (Leal Filho 2025). These capabilities can support more informed and responsive design processes, particularly in contexts where environmental variables and performance criteria are dynamic and interdependent. However, AI also presents significant epistemological challenges, including the opacity of algorithmic processes, biases embedded in data collection and interpretation, and the tendency to prioritise efficiency and utility over ethical, environmental, or socio-cultural considerations. These issues complicate the role of AI in sustainable design. Moreover, AI's influence extends beyond automation, shaping how sustainable design problems are framed, which variables are considered relevant, and what forms of knowledge are valuable in decision-making processes (Ritter et al., 2022). By cultivating both technical proficiency and critical reflection, sustainable design education can utilise AI to explore how technological systems encode, reproduce, or challenge visions of sustainable futures.

Conclusively, sustainable design education is developing into a space where students need to interpret and navigate complex systems involving innovation, environmental sustainability, and socio-cultural transformation. Interdisciplinary and transdisciplinary collaboration emerges as a vital component of curricula as sustainable design continues to intersect with environmental science, artificial intelligence, data analytics, and cultural theory (Keitsch et al., 2020). These developments necessitate new epistemologies that move beyond traditional educational frameworks as well as giving apprentices a more profound understanding of the designer's role and professional responsibilities in relation to diverse stakeholders. Within these developments, environmental ethics appears as reflective practice embedded within curricula rather than providing a fixed set of codes or universal principles. Contemporary sustainable design education conceptualises ethics as a dynamic, context-specific, and process-oriented practice of analysis and decision-making. It encourages students to critically engage with the dialectics of technology, nature and social values, innovation and responsibility, and local and global modes of knowledge generation.

Sustainable design is successively developing more pluralistic (Akama et al., 2016), interconnected and systemic (Culén & Srivastava, 2021), and ethically complex concepts (Fry et al., 2021), and this transition is also visible in a shift from traditional educational goals centred on mastery and problem-solving towards sustainable design curricula that prioritise adaptability, responsiveness, and situated forms of knowing.

Notes

1 Disruption is also part of the "productive paradox", i.e. meaning is generated but cannot be fully mastered, resolved, or optimised. Derrida's "undecidable" (1981) signifies a structurally necessary disruption that reveals the instability of meaning in concepts and decisions. For example: A policymaker must decide whether to subsidise electric cars to mitigate climate change. Reducing CO_2 emissions, e-car batteries cause environmental and social harm. Benefits and harms cannot be fully measured or compared, which leaves undecidable whether electric cars are sustainable, yet a decision is unavoidable. Lyotard (2011) illustrates how the disruptive, figural and undecidable enable and disrupt meaning in terms of artistic production: *"Art stands in alterity as plasticity and desire, a curved expanse against invariability and reason, diacritical space"* (ibid., 7).

2 BioDialectics does not necessarily reject instrumental values of nature but seeks an ethical approach that accepts and includes non-instrumental values in environmental discourses. Respect, autonomy, and non-identity are central ethical values within this perspective. Respect entails acknowledging the inherent worth of both human and nonhuman beings, recognising their right to exist and flourish beyond their utility. It complements instrumental views by cultivating attentiveness and humility. Autonomy is understood heuristically as the capacity for self-determination and resistance to domination, challenging anthropocentric control and affirming reciprocal modes of coexistence. Non-identity resists reductive classifications, instead emphasising the irreducible complexity and alterity of beings. In BioDialectics, it signals an ethical orientation toward what remains beyond human comprehension or assimilation.

3 Environmental awareness, in this context, refers not only to cognitive recognition of ecological issues but to a deeper ethical, aesthetic, and embodied attentiveness to the more-than-human world. Ethically, it involves a sensitivity to the interdependence of life forms and a responsibility toward sustaining shared environments. Aesthetically, it emerges through sensory and affective engagement with nature, where perception fosters a relational understanding (Brady & Prior, 2020).

4 "Productive paradox" comprises antinomies that, rather than obstructing thought or action, contribute to generating new insight, innovative concepts, or artworks. The term characterises a tension that is not resolved but fuels philosophical, ethical, or artistic-productive development. The idea is inspired by Adorno's *Negative Dialectics* (2001) and Morton's *The Ecological Thought* (2010): *"Nature elegy is a paradox, as it's about losing something we never really had: losing a fantasy, not a reality … The ecological thought demonstrates that the aesthetic dimension is full of emptiness-gaps and openness-rather than being a solid, plastic thing. It has no authority. So ecological art is an art of 'whateverness'"* (ibid.,105).

5 The Iban are an Indigenous ethnic group in Sarawak, Malaysia, as well as in parts of Brunei and Kalimantan (Indonesia) (Sather, 1993).

6 Freire's (1970) pedagogy is grounded in critical consciousness, dialogue, and the co-construction of knowledge. He emphasises a participatory, reflective learning process where learners engage with social, political, and environmental realities to

transform them. Rigid knowledge hierarchies are questioned, and education is positioned as a liberating practice aimed at empowerment and social transformation.

7 Which might be specifically relevant for community collaboration and participatory design as *opacity* (e.g. with Glissant) can be interpreted here as a right to difference and non-explainability, emphasising cultural complexity and the autonomy of the Other.

References

Adorno, T. W. (1997). *Aesthetic theory.* Trans. R. Hullot-Kentor. Continuum.

Adorno, T. W. (2001). *Negative dialectics.* Trans. Dennis Redmond in 2021 with corrections. https://platypus1917.org/wp-content/uploads/Negative_Dialectics_Redmondtrans2021.pdf. Retrieved 20 May 2025.

Akama, Y., & Yee, J. (2016). Seeking stronger pluralism in design education. *Design Issues*, 32(1), 31–43.

Appadurai, A. (2013). *The future as cultural fact: Essays on the global condition.* Verso.

Araki, M. (2021). Four components of expertise. Available at SSRN: https://ssrn.com/abstract=3806101 or doi:10.2139/ssrn.3806101. Retrieved 20 May 2025.

Ariansen, P. (1997). The non-utility value of nature – a contribution to understanding the value of biological diversity. Communications of the Norwegian Forest Research Institute. Ås.

Barrett, B. F. D., Horne, R., & Fien, J. (2016). The ethical city: A rationale for an urgent new urban agenda. *Sustainability*, 8(11), 119.

Baskerville, R. L., Kaul, M., & Storey, V. C. (2015). Genres of inquiry in design-science research: Justification and evaluation of knowledge production. *MIS Quarterly*, 39(3), 541–564.

Bennett, J. (2016). *Vibrant matter: A political ecology of things.* Duke University Press.

Bergum, V., & Dossetor, J. B. (2005). *Relational Ethics: The full meaning of respect.* Hagerstown, MD: University Publications Group.

Borthwick, M., Tomitsch, M., & Gaughwin, M. (2022). From human-centred to life-centred design: Considering environmental and ethical concerns in the design of interactive products. *Journal of Responsible Technology*, 10. https://www.sciencedirect.com/science/article/pii/S2666659622000099. Retrieved 28 February 2025.

Brady, E., & Prior, J. (2020). Environmental aesthetics: A synthetic review. *People and Nature*, 2(5), 1189–1206.

Camocini, B., & Vergani, F. (Eds.) (2021). *From human-centered to more-than-human design: Exploring the transition.* Franco Angeli. https://library.oapen.org/handle/20.500.12657/53634. Retrieved 28 February 2025.

Chen, C. (1957). Aristotle's concept of primary substance in books Z and H of the Metaphysics. *Phronesis*, 2(1), 46–59.

Collet, C. (2020). *Designing with living organisms: A biodesign education frame-work. International Journal of Art & Design Education*, 39(1).

Cronon, W. (1996). The trouble with wilderness: Or, getting back to the wrong nature. In W. Cronon (Ed.), *Uncommon ground: Rethinking the human place in nature* (pp. 69–90). W. W. Norton.

Culén, A. L., & Srivastava, S. (2021). Gen MZ toward sustainable fashion practices: Alternative services and business models. In P. Kommers & M. Macedo (Eds.), *Proceedings of the IADIS International Conference ICT, Society and Human Beings; Web Based Communities and Social Media 2021; and e-Health 2021 (pp. 61–70).* IADIS Press.

Derrida, J. (1981). *Positions* (A. Bass, Trans.). University of Chicago Press.
Dieleman, H. (2008). Sustainability, art and reflexivity. *Sustainability: A new frontier for the arts and cultures*, 108, 146.
DiSalvo, C., Lodato, T., Jenkins, T., & Rosner, D. (2022). Doing things with things: Design experiments in sustainability and the anthropocene. *Design Issues*, 38(1), 24–36.
Fernandes, J. P. A., & Guiomar, N. (2016). Environmental ethics: Driving factors beneath behavior, discourse and decision-making. *Journal of Agricultural and Environmental Ethics*, 29(4), 637–656.
Freire, P. (1970). *Pedagogy of the oppressed.* Continuum.
Fry, T., & Nocek, A. (2021). *Design in crisis: New worlds, philosophies and practices.* Routledge.
Gatto, G., & McCardle, J. (2019). Multispecies design and ethnographic practice: Following other-than-humans as a mode of exploring environmental issues. *Sustainability*, 11 (18). https://www.mdpi.com/2071-1050/11/18/5032.Retrieved 28 February 2025.
Giaccardi, E., & Redström, J. (2020). Technology and more-than-human design. *Design Issues* 36(4), 33–44.
Gibbons, L. (2020). Regenerative – the new sustainable? *Sustainability*, 12(13) MDPI, https://www.mdpi.com/2071-1050/12/13/5483. Retrieved 28 February 2025.
Glissant, É. (1997). *Poetics of relation* (Trans. B. Wing). University of Michigan Press.
Glissant, É. (2005). *Kultur und Identität.* Wunderhorn Heidelberg, trans. LaPorte (2007) as *Introduction to a poetics of diversity.* https://repository.lsu.edu/cgi/view content.cgi?article=1004&context=honors_etd. Retrieved 20 February 2025.
Guthrie, Stewart (1993). *Faces in the clouds: A new theory of religion.* Oxford University Press.
Habermas, J. (1987) *Theory of communicative action. Volume two: Lifeworld and system: A critique of functionalist reason*. Trans. Thomas A. McCarthy. Beacon Press.
Herling, R. W. (2000). Operational definitions of expertise and competence. *Advances in Developing Human Resources*, 2(1), 8–21.
Hermans, H. (1988). On the integration of nomothetic and idiographic research methods in the study of personal meaning. *Journal of Personality*, 56(4), 785–812.
Hume, D. (1739). *A treatise of human nature*, Book III, Part I, Section I. Online Library of Liberty. Retrieved 20 February 2025.
Hupkes, T., & Hedman, A. (2022). Shifting towards non-anthropocentrism: In dialogue with speculative design futures. *Futures*, 140, 102950.
Iared, V., Torres de Oliveira, H., & Payne, P. (2016). The aesthetic experience of nature and hermeneutic phenomenology. *The Journal of Environmental Education*, 47(3).
Ingold, T. (2013). *Making: Anthropology, archaeology, art and architecture.* Routledge.
Jacobs, A., Anthoni, E., Busseniers, E., Claes, S., Huybrechts, L., van Leemput, M., Vrancken, K., Carriera, L., Colson, N., Dorn, C., Hostetler, A., Janssens, A., Kabendela, G., Kose, A., Luo, D., Majogoro, M., Sanchez, L. R. J., Stepanovic, J., Swillen, A., & Hannes, K. (2025). Co-designing multispecies speculations through biofuturing. *Qualitative Inquiry*, 31(2), 122–135.
Keitsch, M. (2003). Naturästhetik und ökologische Ethik, Eine Einführung. Bioethiana, Forschungsergebnisse [research results] zur Philosophie, Hamburg 2003: Verlag Dr. Kovac.
Keitsch, M., &Vermeulen, W. J. V. (Eds.). (2020). *Transdisciplinarity for sustainability: Aligning diverse practices* (1st ed.). Routledge.
Kinunen, V., & Valtonen, A. (Eds.) (2017). *Living ethics in a more-than-human world.* University of Lapland. (Rovaniemi, Finland).

Koskinen, I., Zimmerman, J., Binder, T., Redström, J., & Wensveen, S. (2011). *Design research through practice: From the lab, field, and showroom*. Morgan Kaufmann.

Küpers, W. (2020). From the Anthropocene to an "Ecocene": Eco-phenomenological perspectives on embodied, anthrodecentric transformations towards enlivening practices of organising sustainably. *Sustainability*, 12(9). MDPI, https://www.mdpi.com/2071-1050/12/9/3633. Retrieved 20 February 2025.

Lakatos, I. (1970). Falsification and the methodology of scientific research programmes. In I. Lakatos & A. Musgrave (Eds.), *Criticism and the growth of knowledge*. Cambridge University Press.

Leal Filho, W., Kim, E., Borsatto, J. M. L. S., & Marcolin, C. B. (2025). Using artificial intelligence in sustainability teaching and learning. *Environmental Sciences Europe, 37*(1), 124.

Louson, E. M. (2018). Never before seen: Spectacle, staging, and story in wildlife film's blue-chip renaissance. Dissertation. University of Toronto, Ontario.

Lyotard, J.-F. (2011). *Discourse, figure* (A. Hudek & M. Lydon, Trans.). University of Minnesota Press.

Maldonado-Torres, N., Bañales, X., Cespedes, K., Esquivel, M., & Sandoval, T. (2023). *Decolonial pedagogy against the coloniality of justice: Educational theory*. Wiley Online Library.

Mang, P., & Reed, B. (2012). *Regenerative development and design: A framework for evolving sustainability*. Wiley.

Makrakis, V., & Kostoulas-Makrakis, N. (2021). Responsibility and co-responsibility in light of COVID-19 and education for sustainability through an Aristotelian lens. *Sustainability and Climate Change*, 14(2), 61–71.

Mattos, C., & Moreira, P. M. C. (2020). *Frugal innovation: A global research companion*. Routledge.

Moore, B. (2008) *Ecology and literature: Ecocentric personification from antiquity to the twenty-first century*, Palgrave Macmillan.

Morton, T. (2010). *The ecological thought*. Harvard University Press.

Myers, W. (2012). *Bio Design: Nature + science + creativity*. Thames & Hudson.

Nhemachena, A. (2024). Curriculum transformation for the futuristic worlds: Design anthropology for twenty-first-century African universities. *Anthropological Forum*, 34(3), 326–347.

Or, Y. (2021). Regenerative practice as transformative design framework. Pivot 2021 Virtual Conference. OCAD University, Toronto.

Osborne, N., & Grant-Smith, D. (2015). Supporting mindful planners in a mindless system: Limitations to the emotional turn in planning practice. *Town Planning Review*, 86(6), 673–695.

Østern, T. P., Jusslin, S., Nødtvedt Knudsen, K., Maapalo, P., & Bjørkøy, I. (2023). A performative paradigm for post-qualitative inquiry. *Qualitative research*, 23(2), 272–289.

Paolini, A, Bhalla, I., & Loring, P. (2024). Doing more good: Exploring the multidisciplinary landscape of regeneration as a boundary concept for paradigm change. *Journal of Agriculture, Food Systems, and Community Development*, 13, 1–16.

Parisi, S., & Hartmann, M. (2022). Materials as agents: A design-oriented perspective on materiality. *Design Studies*, 79.

Raworth, K. (2017). *Doughnut economics: Seven ways to think like a 21st-century economist*. Chelsea Green Publishing.

Reed, B. (2007). Shifting from "sustainability" to regeneration. *Building Research & Information*, 35(6), 674–680.

Ritter, A., & Ferguson, S. (2022). Artificial intelligence in design education: Possibilities and cautions. *International Journal of Art & Design Education*, 41(1), 90–104.

Roudavski, S., & McCormack, J. (2016). Post-anthropocentric creativity. *Digital Creativity*, 27. 3–6. DOI: doi:10.1080/14626268.2016.1151442.

Sanders, E., & Stappers, P. (2008). Co-creation and the new landscapes of design. *CoDesign*, 4(1), 5–18.

Sather, C. (1993). *The Iban of Sarawak: Chronicle of a vanishing world.* Natural History Publications (Borneo) and British Museum Press.

Silva, T. M. M., Ribeiro, M. F. F., & da Silveira Nascimento, S. (2025). Anthropomorphism: As evolutionary byproduct. In T. K. Shackelford (Ed.), *Encyclopedia of religious psychology and behavior.* Springer.

Simon, H. A. (1996). *The sciences of the artificial* (3rd ed.). MIT Press.

Singh, B., & Keitsch, M. (2025). Reflective agency: The practitioner's role in value-informed design practice. *The International Journal of Design Education*, 20(1), 239–264.

Sperb, J. (2016). From nihilism to nostalgia: Blackfish (2013) and the contradictions of the nature documentary. *Journal of Popular Film & Television*, Taylor and Francis.

Tlostanova, M., & Mignolo, W. (2012). *Learning to unlearn: Decolonial reflections from Eurasia and the Americas.* Ohio State University Press.

Tsing, A. L. (2015). *The mushroom at the end of the world: On the possibility of life in capitalist ruins.* Princeton University Press.

Watson, J. (2019). *Lo–TEK: Design by radical indigenism.* Taschen.

Whyte, K. (2017). Indigenous climate change studies: Indigenizing futures, decolonizing the Anthropocene. *English Language Notes*, 55, 153–162.

Wilkie, A., & Michael, M. (2023). The aesthetics of more-than-human design: Speculative energy briefs for the Chthulucene. *Human–Computer Interaction*, 40, 104–116.

Woodhouse, A., & Brisco, R. (2018). Pareidolia: Characterising facial anthropomorphism and its implications for product design. *Journal of Design Research*, 16(2), 83.

York, T., & Becker, C. (2012). Cultivating an ethic of environmental sustainability: Integrating insights from Aristotelian virtue ethics and pragmatist cognitive development theory. *Journal of College and Character*, 13(4), 1–10.

Zamenopoulos, T., & Alexiou, K. (2018). Co-design as a socially active process. *The Journal of Design, Economics, and Innovation*, 4(1), 5–22.

Zavoleas, Y., Stevens, P. R., Johnstone, J., & Davidová, M. (2023). More-than-human perspectives in indigenous cultures: Holistic systems informing computational models in architecture, urban and landscape design towards the post-anthropocene epoch. *Buildings*, 13(1). MDPI, https://www.mdpi.com/2075-5309/13/1/236. Retrieved 20 February 2025.

6 Conclusions and Insights

This book set out to collect and critically examine ideas, methods, and practices in sustainable design and explore how these intersect with environmental ethics. The chapters present a wide range of approaches, each addressing these issues from a distinct perspective. Writing and editing this book has been a demanding and complex endeavour, requiring sustained critical engagement with a wide array of interconnected topics from human–nature relationships to questions of long-term resilience and shared well-being. It involved navigating diverse perspectives and synthesising ideas across disciplines to explore the ethical and philosophical foundations of sustainable design. I have confronted questions about the role and moral obligations of designers, the role of products and materials, and the need to include justice, equity, and responsibility in sustainable design education. What emerges from the chapters is a recognition that sustainable design is a reflexive, ethical, and creative endeavour, essentially requiring a balance between acknowledging nature, fostering innovation, and maintaining human accountability. A recurring theme is that while ethical values must guide decisions, sustainable design concepts must also respond to real-world settings.

Chapter 1 introduces foundational concepts of Environmental Ethics (EE) and Sustainable Design (SD), tracing their evolution and interconnections. Against the backdrop of escalating ecological crises such as climate change, biodiversity loss, and pollution, the chapter emphasises the urgency of ethical responses in both philosophical and design contexts. Environmental ethics emerged as a distinct philosophical discipline in the 1970s, focusing on the moral obligations humans have toward the natural world. Its two dominant ethical positions were biocentrism and anthropocentrism. Biocentrism holds that all living beings possess intrinsic value and are hence deserving of moral consideration in their own right. In contrast, anthropocentrism regards nature primarily through the lens of human advantage, valuing it for the services and resources it provides. While biocentrism emphasises the preservation of nature for its own sake, anthropocentrism supports preservation primarily in the interest of humans.

The chapter further elaborates on the development of sustainable design, which, although originally rooted in commercial and industrial practices, has

DOI: 10.4324/9781003384090-6

gradually evolved into a discipline increasingly attuned to social and environmental responsibilities. Philosophical trajectories reflect how societies relate to and justify their interactions with nature. The chapter continues by exploring historical perceptions of nature, from Aristotle's view of a purposeful cosmos, through medieval Christian theology that regarded nature as divine creation, to the Enlightenment's rationalist and mechanistic perspective, which enabled technological domination and mass consumerism, ultimately contributing to the environmental crisis. The chapter closes by outlining the book's line of inquiry, which adopts a rhizomatic approach that acknowledges the complexity and diversity of the topics discussed. Emphasising the need for transdisciplinary, reflective, and participatory sustainable design processes that integrate ethical reasoning with real-world application, the chapter concludes by presenting BioDialectics, a novel environmental ethics interpretation for sustainable design, acknowledging both human and non-human interests.

Chapter 2 presents a detailed and nuanced exploration of environmental ethics and sustainable design, contextualising theoretical underpinnings and practical implications. The chapter frames a philosophical baseline on how moral reasoning influences both the ethical treatment of the environment and the responsibilities of design professionals in fostering sustainability. It begins with a discussion of moral philosophies that underpin environmental ethics, focusing on two key perspectives: consequentialist and deontological ethics. Consequentialist views, such as utilitarianism and eudaimonism, evaluate actions based on their outcomes. In environmental ethics, utilitarianism has been expanded to consider animal suffering and ecological impacts, while eudaimonism emphasises a fulfilling life linked to virtues like responsibility and justice. Deontological ethics, on the other hand, prioritises duty and motivation over outcomes and sees ethics as an obligation based on rational principles or universal rights. Applied to environmental concerns, deontological positions argue that nature should be respected not because of its utility but because of its intrinsic worth. This idea is echoed in biocentric concepts that call for a moral duty to protect all forms of life, even when they lack human-like reasoning or interests.

The evolution of sustainable design is further mapped through key figures such as Papanek, who challenged the commercial and industrial roots of design and laid the groundwork for critiques of capitalist design ideologies and inspiring movements in social and ecological design. Papanek's work reflects early efforts to merge environmental ethics with practical design strategies. In the late 20th century, sustainable design underwent a technological shift. Concepts such as Design for the Environment (DfE) and Life Cycle Assessment (LCA) emerged, aiming to reduce environmental impacts across a product's lifespan. While these approaches improved resource and energy use as well as material efficiency, their technocratic and instrumental focus drew criticism, ultimately prompting the emergence of socially and culturally oriented sustainable design concepts. These include behavioural interventions, such as Design for Sustainable Behaviour (DfSB), and participatory models that promote social justice, well-being, and community empowerment. DfSB often relies on nudging or persuasive strategies to guide users toward sustainable practices, raising ethical concerns regarding autonomy and manipulation. Emancipatory design models offer a more democratic and inclusive approach. They focus on enabling local, creative communities and fostering sustainable lifestyles through collaborative design processes.

Since the 1980s, sustainable design has increasingly framed nature as both a functional model and a metaphor. Model-oriented perspectives see ecosystems as examples to emulate, mimicking principles such as balance, resilience, and circularity. In contrast, metaphorical interpretations frame nature as a teacher, offering emotional and spiritual insights that inspire ethical awareness. Both views reflect the growing complexity of sustainable design, underlining the designer's dual role as problem solver and moral agent.

In conclusion the chapter explores Deep Ecology, a philosophical current that advocates for the intrinsic value of all living beings and suggests that environmental ethics emerge from a deep, experiential connection with nature. Its principles, such as biophilia and ontological ethics, have inspired various design movements, but it also faces criticism for potentially undermining human autonomy and democratic processes. Ultimately, Chapter 2 argues that sustainable design ethics must synthesise rational, emotional, social, and ecological dimensions, and proposes BioDialectics as an ethical concept adaptable to contemporary sustainable design. While earlier biocentric models have been criticised for anthropomorphising nature or overlooking human cultural contexts, BioDialectics recognises that all value is ultimately ascribed by humans. However, experiential and emotional dimensions of human–nature relationships are identified as having the potential to transcend anthropocentric perceptions. BioDialectics does not claim a direct causality between nature experience and ethical behaviour, but suggests a strong correlation that might motivate designers to engage with ethics through a deepened understanding of human–nature relationships.

Chapter 3 delves into recent ethical developments in sustainable design and their intersection with environmental ethics, emphasising increasingly complex challenges and philosophical, methodological, and practical responses to them. The chapter explores topics such as sustainability transitions, participatory design, aesthetic experience, posthumanism, and speculative design, each offering insights into how design can ethically engage with environmental and societal transformations. Sustainability transitions (ST) is a central concept discussed in this chapter. ST involves complex, systemic processes aimed at structural transformation in society to address issues like climate change and biodiversity loss. Advocating for integrated changes in socio-technical, ecological, and economic systems, some ST concepts also recognise that capitalist ideologies and consumption patterns are major obstacles. Ethical debates in ST often focus on justice, inclusivity, and the need to critically assess power dynamics and stakeholder influence. Some approaches are criticised for being technocratic or ideologically biased, masking neoliberal agendas under the guise of environmental progress. Within this context, "just transition" theories have emerged that emphasise equity, community empowerment, and the social dimensions of sustainability, highlighting the need for ethical frameworks rooted in lived values and diverse, pluralistic lifeworlds.

Another significant discussion relates to participatory design (PD), which places users and communities at the centre of the design process. PD fosters social sustainability by involving diverse stakeholders from problem definition through to solution implementation. The chapter argues that participatory processes enhance inclusivity, trust, and shared ownership of design outcomes. However, it also cautions against tokenistic participation and the ways in which power dynamics can undermine these values. It emphasises the importance of fostering meaningful, emotionally resonant engagement particularly through storytelling, narrative design, and aesthetic experience, to cultivate ethical awareness among both marginalised and mainstream groups. At this point in the chapter, aesthetic experience emerges as a critical link between sustainable design and environmental ethics. Design is not merely a functional or technological endeavour but also an expressive and emotional one. Aesthetic participation, including encounters with nature, art, and designed objects, can evoke emotional, moral, and even spiritual responses that motivate ethical reflection and action. This view is supported by critical theory, particularly Adorno's aesthetic philosophy, which highlights the potential of autonomous artworks to resist instrumental rationality and capitalist commodification. Aesthetic experiences are framed as avenues to explore alternative ways of being, perceive the non-identical, and challenge normative assumptions about the human–nature relationship.

The chapter then explores postmodern and posthuman perspectives, framing them as both critical responses to and developments beyond traditional environmental ethics. Postmodernism encourages a critical examination of how sustainability is defined and practised. While often criticised for its lack of normative grounding, postmodernism allows for the reimagining of sustainability through pluralistic, culturally situated lenses. This philosophical potential is especially relevant for speculative and critical design (SCD), which uses design fictions and imaginative scenarios to challenge prevailing assumptions and provoke public debate. SCD moves beyond utility to explore the existential and ethical dimensions of design.

In posthumanist and new materialist thought, the chapter presents a deeper ontological and ethical challenge to anthropocentrism. These perspectives dissolve the binary between subject and object, emphasising the vitality and agency of non-human entities, including technologies and natural systems. Concepts like Haraway's "ethics of kinship" and MacCormack's "vitalistic ethics" include ideas that recognise interdependence while challenging human dominance over nature. Overall, Chapter 3 presents an interdisciplinary exploration of how sustainable design can be rethought through contemporary ethical, aesthetic, and philosophical lenses. By integrating environmental ethics with participatory processes, aesthetic experience, and posthumanist thought, the technical and functional dimensions of sustainable design are located within moral and cultural settings to transform human–non-human relationships.

Chapter 4 offers a rich tapestry of narratives that reflect the interplay between sustainable design practices and environmental ethics across diverse geographic, cultural, and institutional settings. The chapter draws upon qualitative data from academic curricula to intentional communities to highlight how values like social justice, community, responsibility, and ecological sensitivity manifest in real-world contexts.

From the sustainable design curricula, the results point to a growing awareness that sustainable design must go beyond technical proficiency and include critical reflection, ethical responsibility, and societal engagement. Some institutions embed sustainability within broader social contexts, emphasising themes such as justice, equity, and activism. By incorporating practitioners into the curriculum, these programmes create a vital bridge between academic learning and professional practice. This connection allows students to engage directly with the complexities of real-world projects, where value conflicts often arise. As a result, students are encouraged to develop a more nuanced understanding of professional ethics and sustainability, particularly in negotiating the delicate balance between political ideologies and the demands of design professionalism.

Other institutions demonstrate that sustainable design education operates through a multi-scalar integration of environmental stewardship, cultural heritage, and socio-cultural ethics. These curricula go beyond teaching ecological responsibility in isolation and seek to embed sustainability within existing urban contexts through community engagement and the preservation of heritage. Initiatives such as community gardens and urban retrofitting serve as practical illustrations of how education and ethics converge in participatory, context-sensitive ways, aiming to balance economic viability, ecological impact, and cultural continuity.

Several intentional communities highlight the ethical foundation of sustainable living through the deliberate cultivation of common activities. They found success largely rooted in the members' early and ongoing commitment to collective engagement, with a strong emphasis on trust, shared values, and inclusivity. Although challenges related to affordability and demographic diversity persist, these communities have made a substantial impact in promoting social cohesion, participatory governance, and environmentally responsible lifestyles. Some communities practice deliberate frugality, privileging simplicity, reuse, and creative expression through recycled materials. Environmental ethics here is embedded in the daily rhythms of communal living, self-governance, and creative labour. Others pursue spiritual integration of environmental ethics, where sustainability is not only a practical imperative but also a sacred duty, while some are more pragmatic in specific efforts, such as transitioning to renewable energy. The ethical lesson of the intentional communal narratives is the fostering of agency, equity, and long-term ecological stewardship.

Across the narratives, some key insights emerge. First, environmental ethics is not monolithic; its meaning and practice vary significantly depending on cultural, geographical, and institutional contexts. What constitutes ethical action in one setting may be interpreted quite differently in another, shaped by local traditions, values, and priorities. Second, sustainable design is inherently relational and context dependent. It does not emerge solely from individual effort but often through collective processes of negotiation, participation, and dialogue. Finally, the use of narrative as a method proves useful because narratives illuminate emotional, ethical, and situational dimensions of sustainability that inform and enrich abstract and quantitative approaches.

Chapter 5 presents a forward-thinking synthesis of novel approaches to sustainable design, centred on the integration of environmental ethics. One of the central discussions in the chapter is the elaboration of BioDialectics, which builds on the recognition of the intrinsic value of all life forms while integrating relational and systemic perspectives. BioDialectics suggests that sustainable design should integrate embodied experiences with nature, sensory perception, reflection, and context-sensitive interpretation into research, education, and practice.

The chapter also explores critical anthropomorphism as a philosophical and methodological tool. Drawing from scholars like Guthrie and Glissant, it suggests that interpreting nature through a human lens, while inherently limited, can be productive when done critically. This approach fosters emotional connection and ethical engagement with the non-human world, while also acknowledging the cultural, political, and epistemic entanglements of design. Glissant's notion of opacity is particularly significant here, as it advocates for the right to irreducible difference and relational existence, challenging Western dualisms and inviting a pluralistic, decolonial approach to landscape and environmental design.

Another novel framework presented is post-anthropocentrism, which challenges the foundational assumption of human centrality in ethics and design. This mindset underpins a range of practices, including regenerative design and living systems design. The chapter further delves into the epistemological tensions between nomothetic (universal, scientific) and idiographic (contextual, experience-based) approaches in sustainable design. While the former dominates current practice through standardised tools like LCA and LEED, it often overlooks cultural and ecological specificity. In contrast, idiographic frameworks emphasise place-based knowledge and co-creation – but these are harder to scale or integrate into policy systems.

The chapter argues for a pluralist and even performative epistemology, where knowledge is enacted through embodied, relational processes that transcend binary divisions between science and experience. In terms of pedagogy, it underscores the importance of embedding ethics as reflective practice in sustainable design education. Rather than prescribing rigid codes, ethics are treated as dynamic and context-dependent, involving dialogue, critical analysis, and stakeholder participation. Teaching tools such as participatory design and ethical analysis are framed as essential for cultivating professional responsibility, especially when addressing complex dilemmas like food consumption or material sourcing.

In conclusion, Chapter 5 presents a robust vision for the future of sustainable design grounded in ethical pluralism, relational thinking, and methodological innovation. Its emphasis on BioDialectics, post-anthropocentric pedagogy, and speculative, participatory practices offers a compelling road map for reimagining the designer's role. Yet, these novel approaches come with notable challenges – requiring systemic shifts in education, interdisciplinary collaboration, and the development of new metrics and mindsets. The future of sustainable design, as envisioned here, lies in its ability to navigate complexity, honour ecological entanglements, and create ethically grounded, context-sensitive solutions.

Writing this book presented considerable challenges, and I am fully aware of its limitations. The primary difficulty stems from the epistemological antinomy between philosophy and design. Environmental ethics is rooted in abstract, logical argumentation, whereas design knowledge is inherently practice-based, iterative, and context-dependent. Bridging this philosophical-pragmatic divide has proven difficult, particularly when attempting to integrate ethical theory into sustainable design without over-reliance on narrative forms of evidence, an approach that may, at times, compromise the theoretical rigor of the book.

Furthermore, my decision not to adopt a fixed normative stance or to endorse specific ethical frameworks may leave educators and practitioners without a clear foundation for decision-making. Some concepts, such as speculative design, offer strong theoretical potential but remain underdeveloped in this book in terms of measurable outcomes, practical replicability, and actionable frameworks.

Finally, while the book ambitiously spans philosophy, design, ecology, ethics, and pedagogy, it does not offer structured, action-oriented, or philosophically coherent pathways for embedding these reflections into the practice of sustainable design. On the other hand, I have left out concepts that are too difficult to integrate into sustainable design even if they are intriguing and revolutionary. For instance, Deleuze and Guattari' s ecological thinking, particularly through the concept of *territory*, seeks to dissolve the traditional distinction between nature and culture. *"The territory is first of all the critical distance between two beings of the same species. Critical distance is a relation based on matters of expression"* (Deleuze and Guattari, 1998, 352). This approach shifts the focus from individual organisms situated within static environments to the dynamic *consistency effects* between overlapping habitats, highlighting processes such as cohabitation, mutual competition, symbiosis, and interdependence. In this view, living beings are not merely in environments; they also constitute environments for one another. However, the dense and highly conceptual nature of their terminology and arguments poses significant challenges even for seasoned philosophers. Moreover, the concise format of this issue cannot fully accommodate the depth and nuance their thought demands.

From my perspective, one of the book's main contributions lies in its philosophical depth offering novel theoretical insights for designers, even if it may be considered "light reading" by professional philosophers. Another strength is its epistemological pluralism, allowing a balance between nomothetic and idiographic knowledge systems. The discussion of recent concepts illustrates that sustainable design is evolving not only in response to ecological issues but also through the rethinking of power, ethics, and epistemology. Emphasis on transdisciplinarity and ethical reflexivity helps position designers as agents of eco-social transformation.

I wrote this book as a reflective guide to reimagining sustainable design through an ethical and philosophical lens. It is intended to support scholars, educators, and practitioners in developing their own perspectives for addressing ecological and socio-cultural challenges grounded in critical insight, moral imagination, and contextual sensitivity.

Reference

Deleuze, G., & Guattari, F. (1998). *A thousand plateaus. Capitalism and schizophrenia*. (Trans. B. Massumi). Continuum.

Index

Note: Locators followed by 'n' refer to notes

For Product Safety Concerns and Information please contact our EU representative GPSR@taylorandfrancis.com
Taylor & Francis Verlag GmbH, Kaufingerstraße 24, 80331 München, Germany

www.ingramcontent.com/pod-product-compliance
Ingram Content Group UK Ltd.
Pitfield, Milton Keynes, MK11 3LW, UK
UKHW020141200926
475702UK00005B/851
* 9 7 8 1 0 3 2 4 6 9 8 6 7 *